Mathematical Analysis: A Very Short Introduction

VERY SHORT INTRODUCTIONS are for anyone wanting a stimulating and accessible way into a new subject. They are written by experts, and have been translated into more than 45 different languages.

The series began in 1995, and now covers a wide variety of topics in every discipline. The VSI library currently contains over 700 volumes—a Very Short Introduction to everything from Psychology and Philosophy of Science to American History and Relativity—and continues to grow in every subject area.

Very Short Introductions available now:

ABOLITIONISM Richard S. Newman
THE ABRAHAMIC RELIGIONS Charles L. Cohen
ACCOUNTING Christopher Nobes
ADDICTION Keith Humphreys
ADOLESCENCE Peter K. Smith
THEODOR W. ADORNO Andrew Bowie
ADVERTISING Winston Fletcher
AERIAL WARFARE Frank Ledwidge
AESTHETICS Bence Nanay
AFRICAN AMERICAN HISTORY Jonathan Scott Holloway
AFRICAN AMERICAN RELIGION Eddie S. Glaude Jr
AFRICAN HISTORY John Parker and Richard Rathbone
AFRICAN POLITICS Ian Taylor
AFRICAN RELIGIONS Jacob K. Olupona
AGEING Nancy A. Pachana
AGNOSTICISM Robin Le Poidevin
AGRICULTURE Paul Brassley and Richard Soffe
ALEXANDER THE GREAT Hugh Bowden
ALGEBRA Peter M. Higgins
AMERICAN BUSINESS HISTORY Walter A. Friedman
AMERICAN CULTURAL HISTORY Eric Avila
AMERICAN FOREIGN RELATIONS Andrew Preston
AMERICAN HISTORY Paul S. Boyer
AMERICAN IMMIGRATION David A. Gerber
AMERICAN INTELLECTUAL HISTORY Jennifer Ratner-Rosenhagen
THE AMERICAN JUDICIAL SYSTEM Charles L. Zelden
AMERICAN LEGAL HISTORY G. Edward White
AMERICAN MILITARY HISTORY Joseph T. Glatthaar
AMERICAN NAVAL HISTORY Craig L. Symonds
AMERICAN POETRY David Caplan
AMERICAN POLITICAL HISTORY Donald Critchlow
AMERICAN POLITICAL PARTIES AND ELECTIONS L. Sandy Maisel
AMERICAN POLITICS Richard M. Valelly
THE AMERICAN PRESIDENCY Charles O. Jones
THE AMERICAN REVOLUTION Robert J. Allison
AMERICAN SLAVERY Heather Andrea Williams
THE AMERICAN SOUTH Charles Reagan Wilson
THE AMERICAN WEST Stephen Aron
AMERICAN WOMEN'S HISTORY Susan Ware
AMPHIBIANS T. S. Kemp
ANAESTHESIA Aidan O'Donnell

ANALYTIC PHILOSOPHY Michael Beaney
ANARCHISM Alex Prichard
ANCIENT ASSYRIA Karen Radner
ANCIENT EGYPT Ian Shaw
ANCIENT EGYPTIAN ART AND ARCHITECTURE Christina Riggs
ANCIENT GREECE Paul Cartledge
ANCIENT GREEK AND ROMAN SCIENCE Liba Taub
THE ANCIENT NEAR EAST Amanda H. Podany
ANCIENT PHILOSOPHY Julia Annas
ANCIENT WARFARE Harry Sidebottom
ANGELS David Albert Jones
ANGLICANISM Mark Chapman
THE ANGLO-SAXON AGE John Blair
ANIMAL BEHAVIOUR Tristram D. Wyatt
THE ANIMAL KINGDOM Peter Holland
ANIMAL RIGHTS David DeGrazia
ANSELM Thomas Williams
THE ANTARCTIC Klaus Dodds
ANTHROPOCENE Erle C. Ellis
ANTISEMITISM Steven Beller
ANXIETY Daniel Freeman and Jason Freeman
THE APOCRYPHAL GOSPELS Paul Foster
APPLIED MATHEMATICS Alain Goriely
THOMAS AQUINAS Fergus Kerr
ARBITRATION Thomas Schultz and Thomas Grant
ARCHAEOLOGY Paul Bahn
ARCHITECTURE Andrew Ballantyne
THE ARCTIC Klaus Dodds and Jamie Woodward
HANNAH ARENDT Dana Villa
ARISTOCRACY William Doyle
ARISTOTLE Jonathan Barnes
ART HISTORY Dana Arnold
ART THEORY Cynthia Freeland
ARTIFICIAL INTELLIGENCE Margaret A. Boden
ASIAN AMERICAN HISTORY Madeline Y. Hsu
ASTROBIOLOGY David C. Catling
ASTROPHYSICS James Binney
ATHEISM Julian Baggini
THE ATMOSPHERE Paul I. Palmer
AUGUSTINE Henry Chadwick
JANE AUSTEN Tom Keymer
AUSTRALIA Kenneth Morgan
AUTISM Uta Frith
AUTOBIOGRAPHY Laura Marcus
THE AVANT GARDE David Cottington
THE AZTECS Davíd Carrasco
BABYLONIA Trevor Bryce
BACTERIA Sebastian G. B. Amyes
BANKING John Goddard and John O. S. Wilson
BARTHES Jonathan Culler
THE BEATS David Sterritt
BEAUTY Roger Scruton
LUDWIG VAN BEETHOVEN Mark Evan Bonds
BEHAVIOURAL ECONOMICS Michelle Baddeley
BESTSELLERS John Sutherland
THE BIBLE John Riches
BIBLICAL ARCHAEOLOGY Eric H. Cline
BIG DATA Dawn E. Holmes
BIOCHEMISTRY Mark Lorch
BIOGEOGRAPHY Mark V. Lomolino
BIOGRAPHY Hermione Lee
BIOMETRICS Michael Fairhurst
ELIZABETH BISHOP Jonathan F. S. Post
BLACK HOLES Katherine Blundell
BLASPHEMY Yvonne Sherwood
BLOOD Chris Cooper
THE BLUES Elijah Wald
THE BODY Chris Shilling
THE BOHEMIANS David Weir
NIELS BOHR J. L. Heilbron
THE BOOK OF COMMON PRAYER Brian Cummings
THE BOOK OF MORMON Terryl Givens
BORDERS Alexander C. Diener and Joshua Hagen
THE BRAIN Michael O'Shea
BRANDING Robert Jones
THE BRICS Andrew F. Cooper
BRITISH CINEMA Charles Barr
THE BRITISH CONSTITUTION Martin Loughlin

THE BRITISH EMPIRE Ashley Jackson
BRITISH POLITICS Tony Wright
BUDDHA Michael Carrithers
BUDDHISM Damien Keown
BUDDHIST ETHICS Damien Keown
BYZANTIUM Peter Sarris
CALVINISM Jon Balserak
ALBERT CAMUS Oliver Gloag
CANADA Donald Wright
CANCER Nicholas James
CAPITALISM James Fulcher
CATHOLICISM Gerald O'Collins
CAUSATION Stephen Mumford and Rani Lill Anjum
THE CELL Terence Allen and Graham Cowling
THE CELTS Barry Cunliffe
CHAOS Leonard Smith
GEOFFREY CHAUCER David Wallace
CHEMISTRY Peter Atkins
CHILD PSYCHOLOGY Usha Goswami
CHILDREN'S LITERATURE Kimberley Reynolds
CHINESE LITERATURE Sabina Knight
CHOICE THEORY Michael Allingham
CHRISTIAN ART Beth Williamson
CHRISTIAN ETHICS D. Stephen Long
CHRISTIANITY Linda Woodhead
CIRCADIAN RHYTHMS Russell Foster and Leon Kreitzman
CITIZENSHIP Richard Bellamy
CITY PLANNING Carl Abbott
CIVIL ENGINEERING David Muir Wood
THE CIVIL RIGHTS MOVEMENT Thomas C. Holt
CLASSICAL LITERATURE William Allan
CLASSICAL MYTHOLOGY Helen Morales
CLASSICS Mary Beard and John Henderson
CLAUSEWITZ Michael Howard
CLIMATE Mark Maslin
CLIMATE CHANGE Mark Maslin
CLINICAL PSYCHOLOGY Susan Llewelyn and Katie Aafjes-van Doorn
COGNITIVE BEHAVIOURAL THERAPY Freda McManus
COGNITIVE NEUROSCIENCE Richard Passingham
THE COLD WAR Robert J. McMahon
COLONIAL AMERICA Alan Taylor
COLONIAL LATIN AMERICAN LITERATURE Rolena Adorno
COMBINATORICS Robin Wilson
COMEDY Matthew Bevis
COMMUNISM Leslie Holmes
COMPARATIVE LITERATURE Ben Hutchinson
COMPETITION AND ANTITRUST LAW Ariel Ezrachi
COMPLEXITY John H. Holland
THE COMPUTER Darrel Ince
COMPUTER SCIENCE Subrata Dasgupta
CONCENTRATION CAMPS Dan Stone
CONDENSED MATTER PHYSICS Ross H. McKenzie
CONFUCIANISM Daniel K. Gardner
THE CONQUISTADORS Matthew Restall and Felipe Fernández-Armesto
CONSCIENCE Paul Strohm
CONSCIOUSNESS Susan Blackmore
CONTEMPORARY ART Julian Stallabrass
CONTEMPORARY FICTION Robert Eaglestone
CONTINENTAL PHILOSOPHY Simon Critchley
COPERNICUS Owen Gingerich
CORAL REEFS Charles Sheppard
CORPORATE SOCIAL RESPONSIBILITY Jeremy Moon
CORRUPTION Leslie Holmes
COSMOLOGY Peter Coles
COUNTRY MUSIC Richard Carlin
CREATIVITY Vlad Glăveanu
CRIME FICTION Richard Bradford
CRIMINAL JUSTICE Julian V. Roberts
CRIMINOLOGY Tim Newburn
CRITICAL THEORY Stephen Eric Bronner
THE CRUSADES Christopher Tyerman
CRYPTOGRAPHY Fred Piper and Sean Murphy
CRYSTALLOGRAPHY A. M. Glazer

THE CULTURAL REVOLUTION
Richard Curt Kraus
DADA AND SURREALISM
David Hopkins
DANTE Peter Hainsworth and
David Robey
DARWIN Jonathan Howard
THE DEAD SEA SCROLLS
Timothy H. Lim
DECADENCE David Weir
DECOLONIZATION Dane Kennedy
DEMENTIA Kathleen Taylor
DEMOCRACY Bernard Crick
DEMOGRAPHY Sarah Harper
DEPRESSION Jan Scott and
Mary Jane Tacchi
DERRIDA Simon Glendinning
DESCARTES Tom Sorell
DESERTS Nick Middleton
DESIGN John Heskett
DEVELOPMENT Ian Goldin
DEVELOPMENTAL BIOLOGY
Lewis Wolpert
THE DEVIL Darren Oldridge
DIASPORA Kevin Kenny
CHARLES DICKENS Jenny Hartley
DICTIONARIES Lynda Mugglestone
DINOSAURS David Norman
DIPLOMATIC HISTORY
Joseph M. Siracusa
DOCUMENTARY FILM
Patricia Aufderheide
DREAMING J. Allan Hobson
DRUGS Les Iversen
DRUIDS Barry Cunliffe
DYNASTY Jeroen Duindam
DYSLEXIA Margaret J. Snowling
EARLY MUSIC Thomas Forrest Kelly
THE EARTH Martin Redfern
EARTH SYSTEM SCIENCE Tim Lenton
ECOLOGY Jaboury Ghazoul
ECONOMICS Partha Dasgupta
EDUCATION Gary Thomas
EGYPTIAN MYTH Geraldine Pinch
EIGHTEENTH-CENTURY BRITAIN
Paul Langford
THE ELEMENTS Philip Ball
EMOTION Dylan Evans
EMPIRE Stephen Howe
EMPLOYMENT LAW David Cabrelli
ENERGY SYSTEMS Nick Jenkins
ENGELS Terrell Carver
ENGINEERING David Blockley
THE ENGLISH LANGUAGE
Simon Horobin
ENGLISH LITERATURE Jonathan Bate
THE ENLIGHTENMENT
John Robertson
ENTREPRENEURSHIP Paul Westhead
and Mike Wright
ENVIRONMENTAL ECONOMICS
Stephen Smith
ENVIRONMENTAL ETHICS
Robin Attfield
ENVIRONMENTAL LAW
Elizabeth Fisher
ENVIRONMENTAL POLITICS
Andrew Dobson
ENZYMES Paul Engel
EPICUREANISM Catherine Wilson
EPIDEMIOLOGY Rodolfo Saracci
ETHICS Simon Blackburn
ETHNOMUSICOLOGY Timothy Rice
THE ETRUSCANS Christopher Smith
EUGENICS Philippa Levine
THE EUROPEAN UNION
Simon Usherwood and John Pinder
EUROPEAN UNION LAW
Anthony Arnull
EVANGELICALISM
John G. Stackhouse Jr.
EVIL Luke Russell
EVOLUTION Brian and
Deborah Charlesworth
EXISTENTIALISM Thomas Flynn
EXPLORATION Stewart A. Weaver
EXTINCTION Paul B. Wignall
THE EYE Michael Land
FAIRY TALE Marina Warner
FAMILY LAW Jonathan Herring
MICHAEL FARADAY
Frank A. J. L. James
FASCISM Kevin Passmore
FASHION Rebecca Arnold
FEDERALISM Mark J. Rozell
and Clyde Wilcox
FEMINISM Margaret Walters
FILM Michael Wood
FILM MUSIC Kathryn Kalinak
FILM NOIR James Naremore

FIRE Andrew C. Scott
THE FIRST WORLD WAR
Michael Howard
FLUID MECHANICS Eric Lauga
FOLK MUSIC Mark Slobin
FOOD John Krebs
FORENSIC PSYCHOLOGY
David Canter
FORENSIC SCIENCE Jim Fraser
FORESTS Jaboury Ghazoul
FOSSILS Keith Thomson
FOUCAULT Gary Gutting
THE FOUNDING FATHERS
R. B. Bernstein
FRACTALS Kenneth Falconer
FREE SPEECH Nigel Warburton
FREE WILL Thomas Pink
FREEMASONRY Andreas Önnerfors
FRENCH LITERATURE John D. Lyons
FRENCH PHILOSOPHY
Stephen Gaukroger and Knox Peden
THE FRENCH REVOLUTION
William Doyle
FREUD Anthony Storr
FUNDAMENTALISM Malise Ruthven
FUNGI Nicholas P. Money
THE FUTURE Jennifer M. Gidley
GALAXIES John Gribbin
GALILEO Stillman Drake
GAME THEORY Ken Binmore
GANDHI Bhikhu Parekh
GARDEN HISTORY Gordon Campbell
GENES Jonathan Slack
GENIUS Andrew Robinson
GENOMICS John Archibald
GEOGRAPHY John Matthews
and David Herbert
GEOLOGY Jan Zalasiewicz
GEOMETRY Maciej Dunajski
GEOPHYSICS William Lowrie
GEOPOLITICS Klaus Dodds
GERMAN LITERATURE Nicholas Boyle
GERMAN PHILOSOPHY
Andrew Bowie
THE GHETTO Bryan Cheyette
GLACIATION David J. A. Evans
GLOBAL CATASTROPHES Bill McGuire
GLOBAL ECONOMIC HISTORY
Robert C. Allen
GLOBAL ISLAM Nile Green
GLOBALIZATION Manfred B. Steger
GOD John Bowker
GÖDEL'S THEOREM A. W. Moore
GOETHE Ritchie Robertson
THE GOTHIC Nick Groom
GOVERNANCE Mark Bevir
GRAVITY Timothy Clifton
THE GREAT DEPRESSION AND THE
NEW DEAL Eric Rauchway
HABEAS CORPUS Amanda L. Tyler
HABERMAS James Gordon Finlayson
THE HABSBURG EMPIRE
Martyn Rady
HAPPINESS Daniel M. Haybron
THE HARLEM RENAISSANCE
Cheryl A. Wall
THE HEBREW BIBLE AS LITERATURE
Tod Linafelt
HEGEL Peter Singer
HEIDEGGER Michael Inwood
THE HELLENISTIC AGE
Peter Thonemann
HEREDITY John Waller
HERMENEUTICS Jens Zimmermann
HERODOTUS Jennifer T. Roberts
HIEROGLYPHS Penelope Wilson
HINDUISM Kim Knott
HISTORY John H. Arnold
THE HISTORY OF ASTRONOMY
Michael Hoskin
THE HISTORY OF CHEMISTRY
William H. Brock
THE HISTORY OF CHILDHOOD
James Marten
THE HISTORY OF CINEMA
Geoffrey Nowell-Smith
THE HISTORY OF COMPUTING
Doron Swade
THE HISTORY OF LIFE
Michael Benton
THE HISTORY OF MATHEMATICS
Jacqueline Stedall
THE HISTORY OF MEDICINE
William Bynum
THE HISTORY OF PHYSICS
J. L. Heilbron
THE HISTORY OF POLITICAL
THOUGHT Richard Whatmore
THE HISTORY OF TIME
Leofranc Holford-Strevens

HIV AND AIDS Alan Whiteside
HOBBES Richard Tuck
HOLLYWOOD Peter Decherney
THE HOLY ROMAN EMPIRE Joachim Whaley
HOME Michael Allen Fox
HOMER Barbara Graziosi
HORMONES Martin Luck
HORROR Darryl Jones
HUMAN ANATOMY Leslie Klenerman
HUMAN EVOLUTION Bernard Wood
HUMAN PHYSIOLOGY Jamie A. Davies
HUMAN RESOURCE MANAGEMENT Adrian Wilkinson
HUMAN RIGHTS Andrew Clapham
HUMANISM Stephen Law
HUME James A. Harris
HUMOUR Noël Carroll
THE ICE AGE Jamie Woodward
IDENTITY Florian Coulmas
IDEOLOGY Michael Freeden
THE IMMUNE SYSTEM Paul Klenerman
INDIAN CINEMA Ashish Rajadhyaksha
INDIAN PHILOSOPHY Sue Hamilton
THE INDUSTRIAL REVOLUTION Robert C. Allen
INFECTIOUS DISEASE Marta L. Wayne and Benjamin M. Bolker
INFINITY Ian Stewart
INFORMATION Luciano Floridi
INNOVATION Mark Dodgson and David Gann
INTELLECTUAL PROPERTY Siva Vaidhyanathan
INTELLIGENCE Ian J. Deary
INTERNATIONAL LAW Vaughan Lowe
INTERNATIONAL MIGRATION Khalid Koser
INTERNATIONAL RELATIONS Christian Reus-Smit
INTERNATIONAL SECURITY Christopher S. Browning
INSECTS Simon Leather
IRAN Ali M. Ansari
ISLAM Malise Ruthven
ISLAMIC HISTORY Adam Silverstein
ISLAMIC LAW Mashood A. Baderin
ISOTOPES Rob Ellam
ITALIAN LITERATURE Peter Hainsworth and David Robey
HENRY JAMES Susan L. Mizruchi
JAPANESE LITERATURE Alan Tansman
JESUS Richard Bauckham
JEWISH HISTORY David N. Myers
JEWISH LITERATURE Ilan Stavans
JOURNALISM Ian Hargreaves
JAMES JOYCE Colin MacCabe
JUDAISM Norman Solomon
JUNG Anthony Stevens
THE JURY Renée Lettow Lerner
KABBALAH Joseph Dan
KAFKA Ritchie Robertson
KANT Roger Scruton
KEYNES Robert Skidelsky
KIERKEGAARD Patrick Gardiner
KNOWLEDGE Jennifer Nagel
THE KORAN Michael Cook
KOREA Michael J. Seth
LAKES Warwick F. Vincent
LANDSCAPE ARCHITECTURE Ian H. Thompson
LANDSCAPES AND GEOMORPHOLOGY Andrew Goudie and Heather Viles
LANGUAGES Stephen R. Anderson
LATE ANTIQUITY Gillian Clark
LAW Raymond Wacks
THE LAWS OF THERMODYNAMICS Peter Atkins
LEADERSHIP Keith Grint
LEARNING Mark Haselgrove
LEIBNIZ Maria Rosa Antognazza
C. S. LEWIS James Como
LIBERALISM Michael Freeden
LIGHT Ian Walmsley
LINCOLN Allen C. Guelzo
LINGUISTICS Peter Matthews
LITERARY THEORY Jonathan Culler
LOCKE John Dunn
LOGIC Graham Priest
LOVE Ronald de Sousa
MARTIN LUTHER Scott H. Hendrix
MACHIAVELLI Quentin Skinner
MADNESS Andrew Scull
MAGIC Owen Davies

MAGNA CARTA Nicholas Vincent
MAGNETISM Stephen Blundell
MALTHUS Donald Winch
MAMMALS T. S. Kemp
MANAGEMENT John Hendry
NELSON MANDELA Elleke Boehmer
MAO Delia Davin
MARINE BIOLOGY Philip V. Mladenov
MARKETING
Kenneth Le Meunier-FitzHugh
THE MARQUIS DE SADE John Phillips
MARTYRDOM Jolyon Mitchell
MARX Peter Singer
MATERIALS Christopher Hall
MATHEMATICAL ANALYSIS
Richard Earl
MATHEMATICAL FINANCE
Mark H. A. Davis
MATHEMATICS Timothy Gowers
MATTER Geoff Cottrell
THE MAYA Matthew Restall and
Amara Solari
THE MEANING OF LIFE Terry Eagleton
MEASUREMENT David Hand
MEDICAL ETHICS Michael Dunn and
Tony Hope
MEDICAL LAW Charles Foster
MEDIEVAL BRITAIN John Gillingham
and Ralph A. Griffiths
MEDIEVAL LITERATURE
Elaine Treharne
MEDIEVAL PHILOSOPHY
John Marenbon
MEMORY Jonathan K. Foster
METAPHYSICS Stephen Mumford
METHODISM William J. Abraham
THE MEXICAN REVOLUTION
Alan Knight
MICROBIOLOGY Nicholas P. Money
MICROBIOMES Angela E. Douglas
MICROECONOMICS Avinash Dixit
MICROSCOPY Terence Allen
THE MIDDLE AGES Miri Rubin
MILITARY JUSTICE Eugene R. Fidell
MILITARY STRATEGY
Antulio J. Echevarria II
JOHN STUART MILL Gregory Claeys
MINERALS David Vaughan
MIRACLES Yujin Nagasawa
MODERN ARCHITECTURE
Adam Sharr
MODERN ART David Cottington
MODERN BRAZIL Anthony W. Pereira
MODERN CHINA Rana Mitter
MODERN DRAMA
Kirsten E. Shepherd-Barr
MODERN FRANCE
Vanessa R. Schwartz
MODERN INDIA Craig Jeffrey
MODERN IRELAND Senia Pašeta
MODERN ITALY Anna Cento Bull
MODERN JAPAN
Christopher Goto-Jones
MODERN LATIN AMERICAN
LITERATURE
Roberto González Echevarría
MODERN WAR Richard English
MODERNISM Christopher Butler
MOLECULAR BIOLOGY Aysha Divan
and Janice A. Royds
MOLECULES Philip Ball
MONASTICISM Stephen J. Davis
THE MONGOLS Morris Rossabi
MONTAIGNE William M. Hamlin
MOONS David A. Rothery
MORMONISM
Richard Lyman Bushman
MOUNTAINS Martin F. Price
MUHAMMAD Jonathan A. C. Brown
MULTICULTURALISM Ali Rattansi
MULTILINGUALISM John C. Maher
MUSIC Nicholas Cook
MUSIC AND TECHNOLOGY
Mark Katz
MYTH Robert A. Segal
NANOTECHNOLOGY
Philip Moriarty
NAPOLEON David A. Bell
THE NAPOLEONIC WARS
Mike Rapport
NATIONALISM Steven Grosby
NATIVE AMERICAN LITERATURE
Sean Teuton
NAVIGATION Jim Bennett
NAZI GERMANY Jane Caplan
NEGOTIATION Carrie Menkel-Meadow
NEOLIBERALISM Manfred B. Steger
and Ravi K. Roy
NETWORKS Guido Caldarelli and
Michele Catanzaro
THE NEW TESTAMENT
Luke Timothy Johnson

THE NEW TESTAMENT AS LITERATURE Kyle Keefer
NEWTON Robert Iliffe
NIETZSCHE Michael Tanner
NINETEENTH-CENTURY BRITAIN Christopher Harvie and H. C. G. Matthew
THE NORMAN CONQUEST George Garnett
NORTH AMERICAN INDIANS Theda Perdue and Michael D. Green
NORTHERN IRELAND Marc Mulholland
NOTHING Frank Close
NUCLEAR PHYSICS Frank Close
NUCLEAR POWER Maxwell Irvine
NUCLEAR WEAPONS Joseph M. Siracusa
NUMBER THEORY Robin Wilson
NUMBERS Peter M. Higgins
NUTRITION David A. Bender
OBJECTIVITY Stephen Gaukroger
OCEANS Dorrik Stow
THE OLD TESTAMENT Michael D. Coogan
THE ORCHESTRA D. Kern Holoman
ORGANIC CHEMISTRY Graham Patrick
ORGANIZATIONS Mary Jo Hatch
ORGANIZED CRIME Georgios A. Antonopoulos and Georgios Papanicolaou
ORTHODOX CHRISTIANITY A. Edward Siecienski
OVID Llewelyn Morgan
PAGANISM Owen Davies
PAKISTAN Pippa Virdee
THE PALESTINIAN-ISRAELI CONFLICT Martin Bunton
PANDEMICS Christian W. McMillen
PARTICLE PHYSICS Frank Close
PAUL E. P. Sanders
IVAN PAVLOV Daniel P. Todes
PEACE Oliver P. Richmond
PENTECOSTALISM William K. Kay
PERCEPTION Brian Rogers
THE PERIODIC TABLE Eric R. Scerri
PHILOSOPHICAL METHOD Timothy Williamson
PHILOSOPHY Edward Craig
PHILOSOPHY IN THE ISLAMIC WORLD Peter Adamson
PHILOSOPHY OF BIOLOGY Samir Okasha
PHILOSOPHY OF LAW Raymond Wacks
PHILOSOPHY OF MIND Barbara Gail Montero
PHILOSOPHY OF PHYSICS David Wallace
PHILOSOPHY OF SCIENCE Samir Okasha
PHILOSOPHY OF RELIGION Tim Bayne
PHOTOGRAPHY Steve Edwards
PHYSICAL CHEMISTRY Peter Atkins
PHYSICS Sidney Perkowitz
PILGRIMAGE Ian Reader
PLAGUE Paul Slack
PLANETARY SYSTEMS Raymond T. Pierrehumbert
PLANETS David A. Rothery
PLANTS Timothy Walker
PLATE TECTONICS Peter Molnar
PLATO Julia Annas
POETRY Bernard O'Donoghue
POLITICAL PHILOSOPHY David Miller
POLITICS Kenneth Minogue
POLYGAMY Sarah M. S. Pearsall
POPULISM Cas Mudde and Cristóbal Rovira Kaltwasser
POSTCOLONIALISM Robert J. C. Young
POSTMODERNISM Christopher Butler
POSTSTRUCTURALISM Catherine Belsey
POVERTY Philip N. Jefferson
PREHISTORY Chris Gosden
PRESOCRATIC PHILOSOPHY Catherine Osborne
PRIVACY Raymond Wacks
PROBABILITY John Haigh
PROGRESSIVISM Walter Nugent
PROHIBITION W. J. Rorabaugh
PROJECTS Andrew Davies
PROTESTANTISM Mark A. Noll
PSEUDOSCIENCE Michael D. Gordin
PSYCHIATRY Tom Burns
PSYCHOANALYSIS Daniel Pick
PSYCHOLOGY Gillian Butler and Freda McManus

PSYCHOLOGY OF MUSIC Elizabeth Hellmuth Margulis
PSYCHOPATHY Essi Viding
PSYCHOTHERAPY Tom Burns and Eva Burns-Lundgren
PUBLIC ADMINISTRATION Stella Z. Theodoulou and Ravi K. Roy
PUBLIC HEALTH Virginia Berridge
PURITANISM Francis J. Bremer
THE QUAKERS Pink Dandelion
QUANTUM THEORY John Polkinghorne
RACISM Ali Rattansi
RADIOACTIVITY Claudio Tuniz
RASTAFARI Ennis B. Edmonds
READING Belinda Jack
THE REAGAN REVOLUTION Gil Troy
REALITY Jan Westerhoff
RECONSTRUCTION Allen C. Guelzo
THE REFORMATION Peter Marshall
REFUGEES Gil Loescher
RELATIVITY Russell Stannard
RELIGION Thomas A. Tweed
RELIGION IN AMERICA Timothy Beal
THE RENAISSANCE Jerry Brotton
RENAISSANCE ART Geraldine A. Johnson
RENEWABLE ENERGY Nick Jelley
REPTILES T. S. Kemp
REVOLUTIONS Jack A. Goldstone
RHETORIC Richard Toye
RISK Baruch Fischhoff and John Kadvany
RITUAL Barry Stephenson
RIVERS Nick Middleton
ROBOTICS Alan Winfield
ROCKS Jan Zalasiewicz
ROMAN BRITAIN Peter Salway
THE ROMAN EMPIRE Christopher Kelly
THE ROMAN REPUBLIC David M. Gwynn
ROMANTICISM Michael Ferber
ROUSSEAU Robert Wokler
RUSSELL A. C. Grayling
THE RUSSIAN ECONOMY Richard Connolly
RUSSIAN HISTORY Geoffrey Hosking
RUSSIAN LITERATURE Catriona Kelly
THE RUSSIAN REVOLUTION S. A. Smith
SAINTS Simon Yarrow
SAMURAI Michael Wert
SAVANNAS Peter A. Furley
SCEPTICISM Duncan Pritchard
SCHIZOPHRENIA Chris Frith and Eve Johnstone
SCHOPENHAUER Christopher Janaway
SCIENCE AND RELIGION Thomas Dixon and Adam R. Shapiro
SCIENCE FICTION David Seed
THE SCIENTIFIC REVOLUTION Lawrence M. Principe
SCOTLAND Rab Houston
SECULARISM Andrew Copson
SEXUAL SELECTION Marlene Zuk and Leigh W. Simmons
SEXUALITY Véronique Mottier
WILLIAM SHAKESPEARE Stanley Wells
SHAKESPEARE'S COMEDIES Bart van Es
SHAKESPEARE'S SONNETS AND POEMS Jonathan F. S. Post
SHAKESPEARE'S TRAGEDIES Stanley Wells
GEORGE BERNARD SHAW Christopher Wixson
MARY SHELLEY Charlotte Gordon
THE SHORT STORY Andrew Kahn
SIKHISM Eleanor Nesbitt
SILENT FILM Donna Kornhaber
THE SILK ROAD James A. Millward
SLANG Jonathon Green
SLEEP Steven W. Lockley and Russell G. Foster
SMELL Matthew Cobb
ADAM SMITH Christopher J. Berry
SOCIAL AND CULTURAL ANTHROPOLOGY John Monaghan and Peter Just
SOCIAL PSYCHOLOGY Richard J. Crisp
SOCIAL WORK Sally Holland and Jonathan Scourfield
SOCIALISM Michael Newman
SOCIOLINGUISTICS John Edwards
SOCIOLOGY Steve Bruce
SOCRATES C. C. W. Taylor
SOFT MATTER Tom McLeish
SOUND Mike Goldsmith

SOUTHEAST ASIA James R. Rush
THE SOVIET UNION Stephen Lovell
THE SPANISH CIVIL WAR
Helen Graham
SPANISH LITERATURE Jo Labanyi
THE SPARTANS Andrew J. Bayliss
SPINOZA Roger Scruton
SPIRITUALITY Philip Sheldrake
SPORT Mike Cronin
STARS Andrew King
STATISTICS David J. Hand
STEM CELLS Jonathan Slack
STOICISM Brad Inwood
STRUCTURAL ENGINEERING
David Blockley
STUART BRITAIN John Morrill
SUBURBS Carl Abbott
THE SUN Philip Judge
SUPERCONDUCTIVITY
Stephen Blundell
SUPERSTITION Stuart Vyse
SYMMETRY Ian Stewart
SYNAESTHESIA Julia Simner
SYNTHETIC BIOLOGY Jamie A. Davies
SYSTEMS BIOLOGY Eberhard O. Voit
TAXATION Stephen Smith
TEETH Peter S. Ungar
TELESCOPES Geoff Cottrell
TERRORISM Charles Townshend
THEATRE Marvin Carlson
THEOLOGY David F. Ford
THINKING AND REASONING
Jonathan St B. T. Evans
THOUGHT Tim Bayne
TIBETAN BUDDHISM
Matthew T. Kapstein
TIDES David George Bowers and
Emyr Martyn Roberts
TIME Jenann Ismael
TOCQUEVILLE Harvey C. Mansfield
LEO TOLSTOY Liza Knapp
TOPOLOGY Richard Earl
TRAGEDY Adrian Poole
TRANSLATION Matthew Reynolds
THE TREATY OF VERSAILLES
Michael S. Neiberg
TRIGONOMETRY
Glen Van Brummelen
THE TROJAN WAR Eric H. Cline
TRUST Katherine Hawley
THE TUDORS John Guy
TWENTIETH-CENTURY BRITAIN
Kenneth O. Morgan
TYPOGRAPHY Paul Luna
THE UNITED NATIONS
Jussi M. Hanhimäki
UNIVERSITIES AND COLLEGES
David Palfreyman and Paul Temple
THE U.S. CIVIL WAR Louis P. Masur
THE U.S. CONGRESS Donald A. Ritchie
THE U.S. CONSTITUTION
David J. Bodenhamer
THE U.S. SUPREME COURT
Linda Greenhouse
UTILITARIANISM
Katarzyna de Lazari-Radek and
Peter Singer
UTOPIANISM Lyman Tower Sargent
VATICAN II Shaun Blanchard and
Stephen Bullivant
VETERINARY SCIENCE James Yeates
THE VIKINGS Julian D. Richards
VIOLENCE Philip Dwyer
THE VIRGIN MARY
Mary Joan Winn Leith
THE VIRTUES Craig A. Boyd and
Kevin Timpe
VIRUSES Dorothy H. Crawford
VOLCANOES Michael J. Branney and
Jan Zalasiewicz
VOLTAIRE Nicholas Cronk
WAR AND RELIGION Jolyon Mitchell
and Joshua Rey
WAR AND TECHNOLOGY Alex Roland
WATER John Finney
WAVES Mike Goldsmith
WEATHER Storm Dunlop
THE WELFARE STATE David Garland
WITCHCRAFT Malcolm Gaskill
WITTGENSTEIN A. C. Grayling
WORK Stephen Fineman
WORLD MUSIC Philip Bohlman
WORLD MYTHOLOGY David Leeming
THE WORLD TRADE
ORGANIZATION Amrita Narlikar
WORLD WAR II Gerhard L. Weinberg
WRITING AND SCRIPT
Andrew Robinson
ZIONISM Michael Stanislawski
ÉMILE ZOLA Brian Nelson

Richard Earl

MATHEMATICAL ANALYSIS

A Very Short Introduction

Great Clarendon Street, Oxford, OX2 6DP,
United Kingdom

Oxford University Press is a department of the University of Oxford.
It furthers the University's objective of excellence in research, scholarship,
and education by publishing worldwide. Oxford is a registered trade mark of
Oxford University Press in the UK and in certain other countries

Published in the United States of America by Oxford University Press
198 Madison Avenue, New York, NY 10016, United States of America

British Library Cataloguing in Publication Data
Data available

Library of Congress Control Number: 2022950499

ISBN 978–0–19–886891–0

Printed and bound by
CPI Group (UK) Ltd, Croydon, CR0 4YY

For cherished friends:
Gareth, Natalie,
Owen, and Matthew

Contents

Acknowledgements

Thanks go to Jonah Blain, Daniel Claydon, Martin Galpin, Natalie Lane, Patrick McDermott, James Munro, and Endre Süli for their comments on, and help with, draft chapters.

List of illustrations

Chapter 1
Taming infinity

Throughout the history of mathematics, infinity and the infinitesimal—the infinitely large and the infinitely small—have raised difficulties and paradoxes. For mathematicians, the study of infinity arises naturally, but in different ways from how an artist or theologian might contemplate it. An artist might welcome the challenge of representing some paradox of infinity in their work; for the devout, an infinite deity may be central to their faith. But the mathematician seeks to define, and ultimately limit, infinity, as mathematics cannot fully progress without rigorous and careful definitions. Not surprisingly, this has been a tortuous and rocky journey that has taken centuries, if not millennia, to resolve. If a study of the history of mathematics teaches anything, it is an appreciation of the endeavour involving many people, over many generations, to find right ways of thinking about mathematics. And this is especially true in the field of analysis.

Mathematical analysis, to some extent, seeks to rigorously define infinite processes that arise in mathematics, so that logical arguments can be made and theorems proven. I write 'to some extent' because analysis is much more than making previously informal mathematics precise; we will see that analysis has ideas and concepts all of its own, many of which were being studied centuries before there was a notion of analysis as a subject in its own right. We shall also see that the applications of analysis are numerous across mathematics and science; indeed, many of the roots of analysis are a result of humankind's efforts to model the world around us.

Actual and potential infinities

The *counting numbers* begin 1, then 2, then 3, and so on, and are commonly listed as

$$1, 2, 3, \ldots$$

Here the ellipsis '...' signifies that the list goes on forever. We can continue counting without end, never running out of numbers. Each counting number, however large, is itself finite, but we recognize that there are infinitely many counting numbers.

The above captures two different types of infinity. A **potential infinity** is an infinite process that goes on without end. Much of mathematical analysis focuses on such infinite processes. When we say there are infinitely many counting numbers, we are referring to an **actual infinity** as if infinity itself were a number.

Early on, we implicitly meet potential infinities. We learn that numbers can be represented as decimal expansions such as:

$$1/3 = 0.333333333333333\ldots$$
$$1/7 = 0.142857142857142\ldots$$
$$\pi = 3.141592653589793\ldots$$

But what *exactly* does all this mean? What details are hidden by those ellipses? This is definitely a mathematician's question. No experimental scientist or engineer has ever needed to know the accuracy of a value to more than 15 decimal places. The value of π given above is sufficiently accurate for space missions exploring the solar system. So, whilst modern computers have calculated π to trillions of places, there are few benefits of such knowledge in the physical world.

One argument for verifying the first decimal expansion might go as follows. If we write

$$x = 0.333333\ldots$$

and then multiply both sides by 10, we get

$$10x = 3.3333333\ldots$$

(as multiplying by 10 moves the decimal point one place to the right). Subtracting the first equation from the second, we find $9x = 3$ and hence $x = \frac{3}{9} = \frac{1}{3}$ as claimed.

So far as it goes, the above is correct, but not the 'whole truth', until we give a precise definition of what a decimal expansion represents; without that, how can we be certain that the algebraic manipulations above are valid?

The following infinite sum, often named after the Italian mathematician Guido Grandi, gives further evidence of the need for rigour. If we set

$$y = 1 - 1 + 1 - 1 + \cdots,$$

then we might argue that

$$y = (1-1) + (1-1) + (1-1) + \cdots = 0 + 0 + 0 + \cdots = 0$$

or that

$$y = 1 + (-1+1) + (-1+1) + \cdots = 1 + 0 + 0 + \cdots = 1,$$

and so we've shown that $0 = 1$. Without proper definitions, seemingly reasonable algebraic manipulations of an infinite sum can lead to nonsensical conclusions.

The first calculation may not seem like an infinite sum, but the notation $x = 0.3333\ldots$ is just shorthand for the infinite sum

$$x = \frac{3}{10} + \frac{3}{100} + \frac{3}{1000} + \cdots.$$

We clearly need an explicit definition of what an infinite sum means, so we can show that $x = \frac{1}{3}$ but exclude contradictory sums like Grandi's. Let's now look at a different way of approaching the problem.

How is π calculated?

In addition to the previous algebraic methods, geometric methods can also be used to determine infinite sums. A square of side 1, and so area 1, can be divided up in two different ways (Figure 1). For example Figure 1(a) shows that

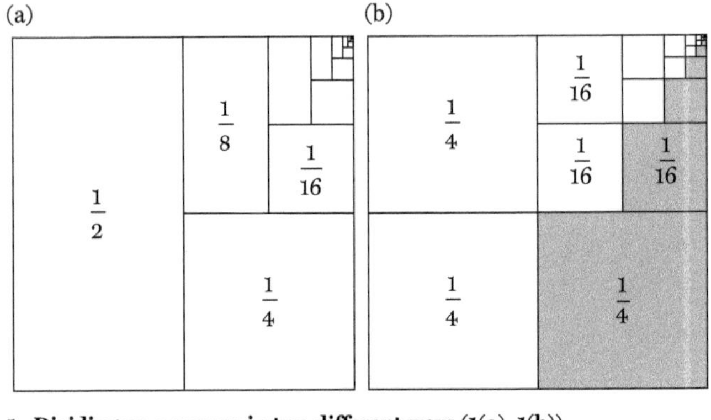

1. Dividing up a square in two different ways (1(a), 1(b)).

$$\frac{1}{2}+\frac{1}{4}+\frac{1}{8}+\frac{1}{16}+\frac{1}{32}+\cdots=1$$

and Figure 1(b) shows that

$$\frac{1}{4}+\frac{1}{16}+\frac{1}{64}+\frac{1}{256}+\frac{1}{1024}+\cdots=\frac{1}{3}.$$

In Figure 1(a), the square is divided up into rectangles and squares, the first region having area $\frac{1}{2}$, and each subsequent region being half the size of the previous region. The squares and rectangles ultimately cover the whole square, and so the infinite sum equals 1. In Figure 1(b), we cover the square with three collections of squares having areas $\frac{1}{4}, \frac{1}{16}, \frac{1}{64}, \ldots$, and so the sum of each collection's areas is one third that of the square, namely $\frac{1}{3}$.

For now, we'll focus on approximating π, which naturally lends itself to geometric methods. Recall that π is the ratio of a circle's circumference to its diameter, and also equals the area of a circle with radius 1.

Here a pentagon has been inscribed in a circle and a hexagon circumscribes the circle (Figure 2). As the pentagon is inside the circle, the pentagon has a smaller area than the circle; as the

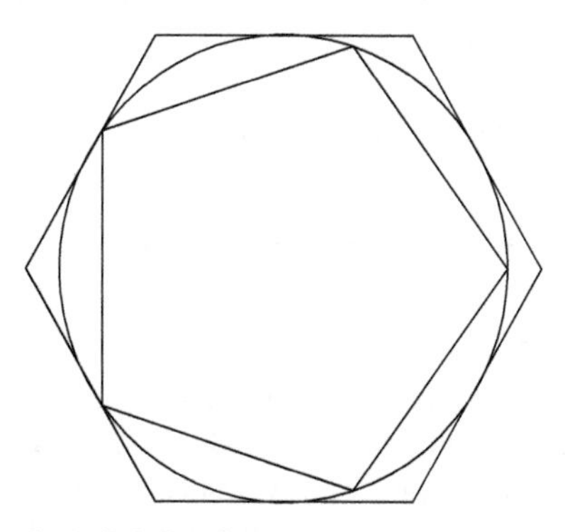

2. A pentagon in a circle in a hexagon.

hexagon surrounds the circle, the hexagon has a larger area. Knowing the polygons' areas would give an underestimate and an overestimate for π. And by using polygons with more edges, these estimates would become progressively better.

In the third century BCE, the great Greek mathematician Archimedes found two approximations of π using such polygons. The two polygons Archimedes used each had 96 edges, and he showed that

$$3.1408\ldots = 3\frac{10}{71} < \pi < 3\frac{1}{7} = 3.1428\ldots,$$

thus determining π to two decimal places. The approximations of π as 3.14 or as $3\frac{1}{7} = \frac{22}{7}$ may already be familiar. This approach was part of a more general *method of exhaustion* that Archimedes employed to great effect, calculating areas by using approximations from inside and outside with regions of known areas.

By the 5th century, Chinese mathematicians had calculated π to seven decimal places using polygons with 24,576 sides. At the start of the 20th century, the number of known places of π was in the hundreds, whilst currently it is in the trillions. Modern approximations of π are achieved analytically using infinite sums, rather than geometrically.

One of the first known expressions for π as an infinite sum is

$$\pi = 4 \times \left(\frac{1}{1} - \frac{1}{3} + \frac{1}{5} - \frac{1}{7} + \frac{1}{9} - \frac{1}{11} + \frac{1}{13} - \cdots\right).$$

This was first derived by the Indian mathematician Madhava in the 14th century, but is also associated with James Gregory and Gottfried Leibniz, who each independently found the sum, albeit three centuries later. Quite what π has to do with the reciprocals of odd numbers may seem unclear at the moment; such infinite sums

Table 1. Approximations for π using Madhava's formula

$s_2 = 2.666666\ldots$	$s_{100} = 3.131592\ldots$	$s_{10000} = 3.141492\ldots$
$s_3 = 3.466666\ldots$	$s_{500} = 3.139592\ldots$	$s_{50000} = 3.141572\ldots$
$s_{10} = 3.041839\ldots$	$s_{1000} = 3.140592\ldots$	$s_{100000} = 3.141582\ldots$
$s_{50} = 3.121594\ldots$	$s_{5000} = 3.141392\ldots$	$s_{500000} = 3.141590\ldots$

are typically evaluated using calculus, and we will explore this in Chapter 2. If we write s_n for the sum of the first n terms of the sum—for example, $s_2 = 4 \times (1 - 1/3) = 8/3$ is the sum of the first two terms—then we generate the above approximations of π (Table 1).

The **partial sums** s_n get ever closer to π but only very slowly. After half a million terms, we have only approximated π to six decimal places. Modern estimates for π use infinite sums that approximate π much more quickly, such as Ramanujan's approximation (see Appendix for details). Its first term,

$$\frac{9801}{2206\sqrt{2}} = 3.1415927300\ldots,$$

is accurate to six decimal places, with each successive term giving a further eight decimal places of accuracy.

Defining convergence

A contentious question, recurring on the internet (pun intended!), is the following:

does 0.99999... equal 1?

The discussions become heated in the absence of any definitions.

'Yes' is the only reasonable one-word answer, but a full answer requires time to explain just what $0.99999\ldots$ means.

Arguments for the negative answer might state that the terms in the infinite sequence

$$0.9, 0.99, 0.999, 0.9999, 0.99999, \ldots$$

are always less than 1 and so never equal 1. This is true, but the notation $0.99999\ldots$ does not represent a sequence; it represents a number which is known as the **limit** of this sequence. However, the above sequence does have an important part in defining what the notation $0.99999\ldots$ means.

The individual terms of the sequence never equal 1, but they do become arbitrarily close to 1. That is, given any required degree of approximation, some term of the sequence gets that close to 1. For example, if we required a term that was within an accuracy of 0.0035 of 1, then we'd need that term to satisfy

$$0.9965 = 1 - 0.0035 < \text{term of sequence} < 1 + 0.0035 = 1.0035.$$

Note that the third term, 0.999, is in this range and, in fact, every term afterwards is in this range as well.

And this is essentially the definition of convergence. A mathematical textbook might introduce more general notation, and a broader setting, but a sequence of numbers converges to a limit if, for any required degree of accuracy, the terms of the sequence eventually become that accurate and remain so.

Returning to Grandi's sum, we can use our definition to show that it doesn't converge. The partial sums are

$$1, \quad 1-1, \quad 1-1+1, \quad 1-1+1-1, \quad 1-1+1-1+1,$$

which evaluate to $1, 0, 1, 0, 1, \ldots$ Half of the partial sums are 1 and half are 0. If we chose a required accuracy of 0.1, then a limit, if it existed, would have to be within 0.1 of 1, so between 0.9 and 1.1, *and* the limit would also have to be within 0.1 of 0, so between -0.1 and 0.1. No number lies in *both* ranges, so Grandi's sum has no limit.

Now that we have a clear definition, these matters of convergence resolve straightforwardly. Well, sort of. What I've described here is the standard definition of convergence for sequences. However, in 1890 the Italian analyst Ernesto Cesàro introduced a more general notion of convergence for which Grandi's sum *does* converge and takes a value of $1/2$. Which is correct? Does Grandi's sum converge or not?

The answer is that they are both right: different definitions can lead to different answers, and to be clear we should state which definition is being used. Otherwise, it's only reasonable to expect that the standard definition of convergence is being used.

By way of a more extreme example, you may have seen the following infinite sum:

$$1 + 2 + 3 + 4 + \cdots = -\frac{1}{12},$$

which is commonly cited by string theorists in physics. This seems wholly ridiculous; the partial sums of $1, 3, 6, 10, \ldots$ increase forever and are never negative. They certainly don't converge by the standard definition, but rather tend to infinity. Cesàro would have agreed. However, in around 1913, Srinivasa Ramanujan introduced an approach which assigned precisely the above value to this infinite sum.

We shall focus only on the standard notion of convergence and consider Cesàro's and Ramanujan's sums too niche for further discussion. I include them here to make clear that context and choice of definition matter; they are important enough that a mathematical question might have different answers depending on how it is understood.

Countable versus uncountable

Potential infinities are processes that go on forever; an important aspect of analysis is the handling of such processes, defining clearly how they might resolve and how two such processes might interact. At this point a reasonable question would be: 'why do we need such processes in the first place?' Large parts of mathematics involve only finite processes. No computer has ever literally added an infinite number of terms together, as that would take forever.

A potential infinity is a process like the counting of $1, 2, 3, \ldots,$ which never ends. An actual infinity is the answer to the question 'how many counting numbers are there?' During the late 19th century, the German mathematician Georg Cantor defined ways to rigorously investigate *actual infinities*.

The beginnings of analysis are intimately linked with the real numbers. A **real number** is any number with a decimal expansion; the set of real numbers includes the counting numbers $1, 2, 3, \ldots,$ the *rational* numbers (these are fractions of whole numbers such as $-\frac{1}{3}$ and $\frac{22}{7}$) and other *irrational* (= not rational) numbers such as $\sqrt{2}$ and π. In 1874 Cantor showed that there are more real numbers than there are counting numbers. That is, it's impossible to count or list all the real numbers. Any attempted listing—first real number, second real number, third real number, . . .—would necessarily omit some real numbers (most of them, in fact). A proof is given in the Appendix. This result speaks to the nature of the real numbers and why analysis needs infinite processes.

A finite set or an infinite set which can be counted is called, unsurprisingly, *countable*; otherwise, it's called *uncountable*. The counting numbers are countable; the integers (positive and negative whole numbers) are countable; perhaps more surprisingly, the rational numbers are countable. Yet more surprisingly, the *computable* numbers are countable.

A computable number is one which a computer can approximate to any required degree. The rational numbers are all computable. Other numbers like π are computable; a computer could be programmed with either Madhava's or Ramanujan's sum to calculate π to any required accuracy. A computer program is just a list of commands of finite length, written in a computer language comprising finitely many symbols. Using Cantor's ideas, it can be shown that there are countably many programs. So there are countably many computable numbers.

On the other hand, there are uncountably many real numbers, when understood as numbers with arbitrary decimal expansions. This means that some (in fact, most) real numbers cannot be described by finite means, and so analysis needs infinite processes and descriptions to deal with the real numbers.

Most mathematicians are fine with this, but for some philosophers and logicians this is a great concern. There are schools of thought, in particular that of the *intuitionists*, which do not accept the notion of arbitrary decimal expansions and so have a different notion of what can be validly proved in analysis, though this is not the broadly held view of mathematicians. Importantly, the uncountability of the real numbers makes clear that finite processes are insufficient to be able to describe them.

Axioms and some early results

Rigorous definitions are important to mathematicians, as we can employ them in rigorous proofs—carefully argued chains of logic that begin with clear assumptions or previously demonstrated results, making clear deductions at each step until we reach a conclusion. In a fully understood subject, definitions come before theory and carefully ground the assumptions and logical deductions made in a proof. Historically, as calculus emerged, the situation was much more chicken-and-egg as understanding progressed, and we will see that rigorous definitions came quite late in the narrative. Ideally though, definitions are the key starting points to proving theorems.

A modern introduction to real analysis would likely begin with some **axioms** of the real numbers; an axiom is an assumed rule that is considered self-evident, and doesn't need to be proved. After all, without making *some* assumptions, nothing can be proved. For example, the *commutativity* of addition states, for any two real numbers x and y, that

$$x + y = y + x.$$

So the order of addition doesn't matter. If we are asked to add 3 and 4, we get a sum of 7, whether we work this out as 3 + 4 or 4 + 3. Other axioms relate to the ordering of the real numbers, and one example is the *transitivity* of order. This states that

if $x < y$ and $y < z$, then $x < z$.

A more subtle assumption is the **completeness axiom**. This states that

an increasing bounded real sequence converges.

An increasing sequence is one in which the next term is always at least as great as the current term, and a bounded sequence is one where all the terms lie between two fixed real numbers. By way of example, the first two sequences from

$$1,\ 2,\ 3,\ 4,\ 5\ldots,$$
$$3,\ 3.1,\ 3.14,\ 3.141,\ 3.1415,\ \ldots,$$
$$0,\ 1,\ 0,\ 1,\ 0,\ \ldots$$

are increasing; the third sequence is not increasing, as the third term of 0 is less than the second term of 1. However, the first sequence does not have an upper bound. The terms of the second sequence are the terminating decimal expansions of π; these terms are bounded above by 4 and below by 3. The completeness axiom states that this sequence converges, which it does, namely to π. (Fuller details about the axioms appear in the Appendix.)

Once we have agreed axioms, we can prove some first analytic results such as the **uniqueness of limits.** We showed earlier that the sequence $0.9,\ 0.99,\ 0.999\ldots$ converges to 1, but did not contemplate whether another real number might also be the limit of the sequence. In fact, this cannot arise: a limit, if it exists, is unique. Other early results include the **algebra of limits.**

Here are two convergent sequences:

$$0.1,\ 0.11,\ 0.111,\ 0.1111,\ldots \quad 0.1,\ 0.18,\ 0.181,\ 0.1818,\ldots$$

The first sequence converges to $\frac{1}{9}$ and the second converges to $\frac{2}{11}$. We can create a new sequence by adding the sequences termwise to get

$$0.2,\ 0.29,\ 0.292,\ 0.2929,\ \ldots$$

It turns out that this sequence converges, and moreover converges to

$$\frac{29}{99}=\frac{1}{9}+\frac{2}{11}.$$

The sum of these two sequences converges to the sum of their limits, and similar results hold when we subtract, multiply, or divide convergent sequences. It is precisely the algebra of limits that makes valid the earlier argument, which showed that $0.3333\ldots = \frac{1}{3}$. By the completeness axiom, we know that $0.3333\ldots$ represents some real number x, and by the algebra of limits we know that

$$3,\ 3.3,\ 3.33,\ 3.333,\ \ldots \text{ converges to } 10x.$$

The difference of the sequences, which is the constant sequence $3,\ 3,\ 3,\ 3,\ \ldots$ converges to the difference of the limits $10x - x = 9x$. By the uniqueness of limits, $9x = 3$ and hence $x = \frac{1}{3}$.

Such positive results are reassuring, but there is still room for some early counter-intuitive results. If we have an infinite sum of positive terms which converges, such as

$$\frac{1}{4} + \frac{1}{16} + \frac{1}{64} + \frac{1}{256} + \frac{1}{1024} + \cdots = \frac{1}{3}$$

(Figure 1(b)), then however we reorder the terms, the sum will still converge to $\frac{1}{3}$. This agrees with our experience with finite sums; from the axioms of the real numbers it can be proved that a finite number of terms give the same result, whatever order they are added. But this need not remain true for infinite sums involving both positive and negative terms. Recall the Madhava sum:

$$0.78539\ldots = \frac{\pi}{4} = \frac{1}{1} - \frac{1}{3} + \frac{1}{5} - \frac{1}{7} + \frac{1}{9} - \frac{1}{11} + \frac{1}{13} - \cdots$$

We can reorder the terms as

$$\frac{1}{1} + \frac{1}{5} - \frac{1}{3} + \frac{1}{9} + \frac{1}{13} - \frac{1}{7} - \cdots,$$

so that the positive terms are summed two at a time compared with the negative terms. Note that all the terms are still present, and none duplicated; the second sum has all the same terms as the first, just in a new order. However, this time the infinite sum can be shown to converge to 0.95868 . . . which is greater. (For those with knowledge of logarithms, the exact sum is $(\pi + \log 2)/4$.) In fact, the German mathematician Bernhard Riemann showed in 1853 that these terms can be rearranged to sum to *any* value, finite or infinite. If we wanted, say, a sum of 100, then we would carefully need to front-load the rearrangement with positive terms to manage this, but we could achieve this, as, importantly, there are infinitely many positive terms, and the sum of those positive terms is infinite.

This may or may not be strikingly counter-intuitive to you, but I hope you agree that some people would find it so. Without careful definitions, without careful proofs, it is impossible to convince others when intuition fails.

Modern analysis

Analysis itself arose as a separate subject within mathematics around the 19th century. We shall see, though, that the terms *analysis* and *analytic* occurred much earlier, particularly in the 17th century with the work of Fermat and Descartes on 'analytic geometry'—that is, co-ordinate geometry.

The 19th century was when modern analytic treatments became recognizable, such as so-called ε-δ (read: 'epsilon-delta') proofs (Chapter 2). The definition of convergence given earlier is commonly attributed to the German analyst Karl Weierstrass, who was lecturing in Berlin on such material in 1861. Weierstrass is often considered the father of modern analysis and remembered for introducing ε-δ arguments. Much earlier, though, in 1817, Bernard Bolzano had made the same definition and proved several important theorems of analysis, but his work did not receive due attention for another 50 years. Augustin-Louis Cauchy also makes use of ε-δ proofs in his influential *Cours d'Analyse* of 1821. So, as with much mathematics, a rigorous treatment of real analysis arose over decades and from the contributions of many.

A modern undergraduate mathematics course on calculus (Chapter 2) includes limits as a central, foundational concept. But by the 19th century calculus was almost two centuries old; it had been widely applied and studied without a formal definition of limit, though that is not to say it had been without its critics. Increasingly, a rigorous approach to analysis, as well as a broadening in the notion of a function, was becoming necessary, especially in the treatment of Fourier series (Chapter 6). A growing appreciation of analytic matters would lead to more than just a firmer understanding of old results. As we shall see in later chapters, analysis would find a canon all of its own, and the theory and methods of analysis would have impact across mathematics and find applications in much of science.

Chapter 2
All change . . . the calculus of Fermat, Newton, and Leibniz

Calculus

The invention of calculus is traditionally credited to Newton and Leibniz in the late 17th century, though their progress was very much built on the work of others. Given the sheer range of applications it has found in mathematics and the physical sciences, calculus can arguably be described as humankind's single greatest invention from the last 500 years. It has, as the mathematical historian C. H. Edwards writes, 'served for three centuries as the principal quantitative language of Western science'. Calculus most naturally falls into two branches: *differential calculus*—the study of rates of change—and *integral calculus*—the study of accumulated changes. These two processes, differentiation and integration, are essentially inverses of one another, a fact made explicit in the *fundamental theorem of calculus.*

Some of the questions calculus addresses date to the ancient Greeks, but much of the early focus of calculus related to contemporary scientific problems, particularly physical and astronomical ones. Within geometry, calculus can determine tangent lines, areas, and volumes; beyond pure mathematics, and particularly via *differential equations*, calculus would help model and understand much of the world around us. As they evolved, the approaches to calculus would take on the flavour of the new algebra and new co-ordinate geometry, much more than that of ancient Greek geometry. We shall see that calculus involves the sorts of limit processes that are now naturally thought of as part of modern analysis. The rise of calculus and its widespread applications would only add to the urgency of developing a rigorous grounding for the subject. Equally, the breadth of application of calculus was just a precursor to the subsequent impact of modern analysis across mathematics and science.

To better appreciate how and why calculus developed, we first need to review the important advances of the previous century.

The 17th century

The 17th century was transformative for European mathematics. Between the ancient Greek era and 1600 there had been important advances: the Hindu–Arabic number system had been introduced, promoted by Fibonacci in his *Liber Abaci* of 1202; cubic (degree 3) and quartic (degree 4) polynomial equations had been solved by Italian mathematicians in the 16th century; François Viète had made important improvements in algebraic notation. But as of 1600 the canon of European mathematics was mostly ancient Greek in both content and emphasis. The advances made by the Kerala school in India, for example in the use of power series, were unknown to European mathematicians of the time.

17th-century European mathematics would advance in various ways: in 1614 John Napier introduced logarithms (Chapter 3); projective geometry developed from the study of perspective in art; and, in a correspondence of 1654, Pierre de Fermat and Blaise Pascal laid down many of the fundamentals of probability.

But as far as the development of calculus is concerned, the two important advances of the 17th century were *co-ordinate geometry*, also known as *analytic geometry*, and the concept of *function*.

Synthetic versus analytic geometry

The word 'synthesis' means a process of thought leading from cause to effects; in this sense the geometry of Euclid is synthetic, with theory carefully being deduced from assumed axioms. This contrasts with 'analysis', which means a process of thought leading from effects to cause—or at least this is the word's original etymology. In the word 'psychoanalysis' we can see that the term has retained its roots—underlying disorders are diagnosed from visible symptoms and behaviours—but we shall see that the meaning of 'analysis' in mathematics has evolved considerably since the 17th century.

Back then, the phrase 'analytic geometry' referred to co-ordinate geometry. The methods of analytic geometry introduce undetermined quantities, such as x and y, and the geometric constraints on these unknown quantities manifest as equations involving x and y which are to be solved. At that time, the term analysis was largely synonymous with such use of algebra and equations.

By way of contrast, we give three proofs of Thales' theorem, which states that the angle made by a diameter in a semi-circle is a right angle. They are included to show how different the language of the mathematics is, so don't be concerned if some of the reasoning is unfamiliar.

Each proof begins with a circle, centre O and diameter AB. The third point C of a triangle ABC lies on the circle. We wish to prove that the angle $\angle ACB$ is a right angle (Figure 3).

PROOF 1: Draw in the line OC. As OA and OC are radii, the triangle AOC is isosceles and so the base angles $\angle OAC$ and $\angle OCA$ are equal. Likewise, angles $\angle OBC$ and $\angle OCB$ are equal. Then $2\angle OCA + 2\angle OCB$ equals two right angles, as they make up all the

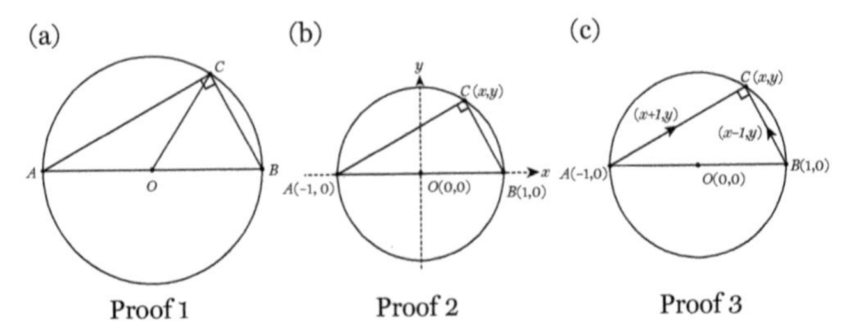

3. Three proofs of Thales' theorem (3(a), 3(b), 3(c)).

angles of the triangle ACB. So finally $\angle ACB = \angle OCA + \angle OCB$ equals one right angle.

PROOF 2: Choose co-ordinates so that $O = (0, 0),\ A = (-1, 0)$, $B = (1, 0), C = (x, y)$; by Pythagoras' theorem, the circle's equation is then $x^2 + y^2 = 1$. The gradient of CB equals $\frac{y}{x-1}$ and the gradient of CA equals $\frac{y}{x+1}$. The product of these gradients equals

$$\frac{y}{x-1} \times \frac{y}{x+1} = \frac{y^2}{x^2-1} = \frac{1-x^2}{x^2-1} = -1$$

and so the lines CB and CA are at right angles.

PROOF 3: With co-ordinates chosen as in Proof 2, consider the vectors $\overrightarrow{AC} = (x+1, y)$ and $\overrightarrow{BC} = (x-1, y)$. Their scalar product equals

$$\overrightarrow{AC} \cdot \overrightarrow{BC} = (x-1)(x+1) + y^2 = x^2 - 1 + y^2 = 0.$$

Hence $\overrightarrow{AC}$ and $\overrightarrow{BC}$ are perpendicular.

The first proof is entirely synthetic; it relies on previous results, such as the base angles of isosceles triangles being equal and the angles of a triangle adding up to two right angles. No mention of co-ordinates is made, the entire problem being set in a featureless Euclidean plane.

The second proof is analytic. It introduces co-ordinates into the Euclidean plane and does so, without any loss of generality, to make the subsequent algebra as palatable as possible. We place the origin on the circle's centre, align the x-axis with the diameter, and take the circle's radius as the unit length. The general point C is assigned co-ordinates (x, y), and all we know is that $x^2 + y^2 = 1$, this being the equation of the circle. The proof relies on the fact that two lines are perpendicular if the product of their gradients equals -1. Note that the concept of gradient doesn't even make sense in synthetic geometry; it's a notion we can only assign to lines once we have introduced co-ordinates. This second proof is closest in style to a 17th-century analytic proof of Thales' theorem.

The third proof has a more modern style which would not have appeared until the 20th century, though it is essentially the same as the second proof. It makes use of vectors and the scalar product.

Analytic geometry and the function concept

Analytic, or co-ordinate, geometry was independently introduced by René Descartes and Fermat; Descartes' work was published in 1637, but Fermat's only appeared posthumously in 1679. Instead of a featureless Euclidean plane, the Cartesian plane—named in honour of Descartes—has two perpendicular *axes*, the horizontal x-axis and the vertical y-axis, meeting at the *origin*. Every point of the Cartesian plane can be uniquely assigned x- and y-co-ordinates, depending on how far along the x- and y-axes the point is (Figure 4(a)). Note that points to the left of the y-axis have a negative x-co-ordinate, so the co-ordinates are displacements from the axes rather than simply distances. (Distances cannot be negative.)

Given the advances the ancient Greeks made in geometry, it is surprising that they made only rudimentary use of co-ordinates, with the arguable exception of Apollonius of Perga, but even he made no use of negative numbers and used co-ordinates to

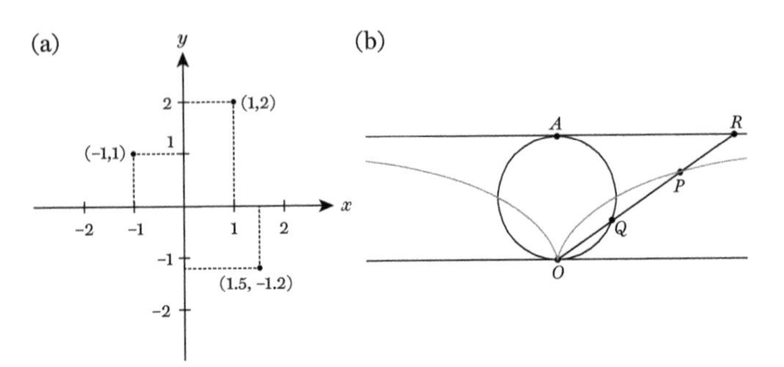

4(a). The Cartesian plane. 4(b). The cissoid of Diocles.

investigate curves that were initially characterized by their geometry.

Let's take the *cissoid of Diocles* as an example (Figure 4(b)). A circle of radius a has base point O which is diametrically opposite the point A. Given a point R on the tangent line to A, the line OR intersects the circle at Q, and a third point P on the line OR is such that the distances OP and QR are equal. The cissoid is the curve traced by P as R moves along the tangent line.

The ancient Greeks investigated curves, like the cissoid, using such geometric constructions. But if we take O to be the origin and position the y-axis along OA, then the cissoid has the equation

$$(x^2 + y^2)y = 2ax^2,$$

meaning a point (x, y) lies on the cissoid precisely when x and y satisfy the above equation. (Details appear in the Appendix.)

Graphs are now so ubiquitous, representing data or some recent trend in the news, that it is hard for us to appreciate the revolutionary impact that the introduction of co-ordinates had on mathematics. But looking at the first and second proofs of Thales' theorem reviewed earlier, the different emphases are marked—one is geometric, one essentially algebraic, each making use of starkly different methods. Also, the second proof implicitly uses advances that might not be clear—an ease with using negative numbers and much improved algebraic notation. Importantly, there was no longer any primacy of geometry over algebra; the cissoid, for example, can be as readily described by the above equation as by its geometric construction. Further, with the introduction of co-ordinates, it's natural to think of points on the cissoid as the graph of some **function** $y(x)$ of a variable x.

It was Leibniz who first coined the term 'function' in 1673. The concept of a function barely existed prior to the 17th century, and its development would continue into the 20th century. With the advent of analytic geometry, it became natural to think of the y-co-ordinate (or 'dependent variable') of a point on a curve as a function of the x-co-ordinate (or 'independent variable'). By modern standards, the description that y might depend on x 'in an algebraic or transcendental manner' (to quote the Swiss mathematician Johann Bernoulli around 1697) is primitive. The development of the function concept and that of analysis would be intimately connected over the next three centuries, with the need for a rigorous and broad notion of a function often driving advances in analysis or vice versa.

Pierre de Fermat

The French mathematician Fermat (Figure 5(a)) is often termed the 'prince of amateurs', as he was actually a lawyer and *parlementaire* in Toulouse by profession. He made significant contributions across much of mathematics and physics: he independently introduced Cartesian co-ordinates, including in three dimensions; in optics the principle that light takes the least time to travel between two points is due to him; and he made many advances in number theory, including proving that every prime number which is one more than a multiple of four can be written as a sum of two squares (e.g. $17 = 1^2 + 4^2, 653 = 13^2 + 22^2$). Fermat, more than

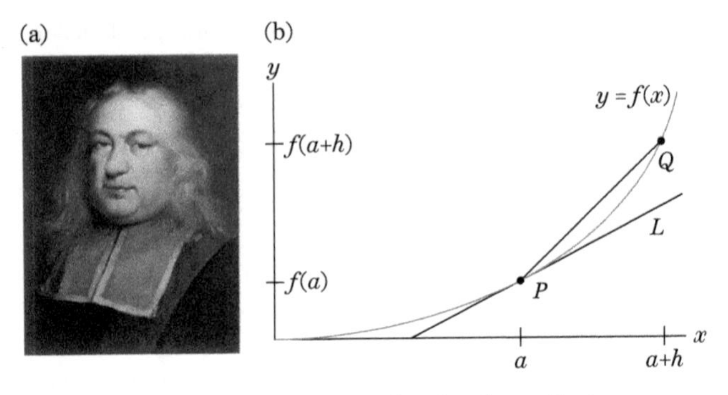

5(a). Pierre de Fermat. 5(b). Approximating the gradient.

any of his contemporaries, showed the power of algebraic, analytic methods when applied to old and new problems; to quote Michael Sean Mahoney from his biography of Fermat, 'in a very real sense, Fermat presided over the death of the classical Greek tradition in mathematics'.

In the development of calculus, Fermat's most important work was his 1636 *Method for Determining Maxima and Minima and Tangents to Curved Lines*. Consider the graph of a function $y = f(x)$ and the tangent line L at a point $P = (a, f(a))$ (Figure 5(b)). This tangent line just touches the graph, having the same gradient as the graph at the point of intersection P.

In order to determine that gradient, Fermat considered a nearby point $Q = (a+h, f(a+h))$ on the graph. The gradient of the chord PQ is the change in the y-co-ordinate divided by the change in the x-co-ordinate, namely:

$$\text{gradient of chord } PQ = \frac{f(a+h) - f(a)}{(a+h) - a} = \frac{f(a+h) - f(a)}{h}.$$

If h is a small non-zero real number, it's reasonable to think that the gradient of PQ will be a good approximation to the tangent's gradient. Note, importantly, we cannot just set h to be zero, as the above fraction would become $\frac{0}{0}$, which is meaningless. We want to say that 'h should become as small as possible'; simple as that phrase seems, it would take mathematicians two centuries to work out quite what they ought to be saying. Fermat himself was never explicit on this matter, introducing a notion of 'adequality' to describe this process, and historians of mathematics continue to discuss quite what he intended by this.

Let's consider a specific choice of function, $f(x) = x^2$. The previous fraction becomes:

$$\frac{(a+h)^2 - a^2}{h} = \frac{a^2 + 2ah + h^2 - a^2}{h} = \frac{2ah + h^2}{h} = 2a + h.$$

Here it seems clear that we get the answer of $2a$, as 'h becomes small'; indeed, it's now valid to set h to equal zero to obtain that answer. And in this case $2a$ is the correct gradient of the graph $y = x^2$ at the point (a, a^2).

More generally, Fermat showed that the graph $y = x^n$ has gradient na^{n-1} when $x = a$. Take special note of this result, as we will repeatedly use it during the rest of the text.

What might not be apparent is that this calculation works out nicely because x^n is a *polynomial* when $n \geqslant 0$—that is, a function of the form

$$f(x) = c_0 + c_1 x + c_2 x^2 + \cdots + c_k x^k,$$

where k is a non-negative whole number and $c_0, c_1, c_2, ..., c_k$ are real numbers. These functions are sufficiently nice that, via some algebraic manipulation, Fermat was left in a position where he could just set h to equal zero. The curves of interest to Fermat were defined by polynomial equations in the co-ordinates x and y, so he never needed to consider how h *tended to zero* or introduce the notion of *limit*.

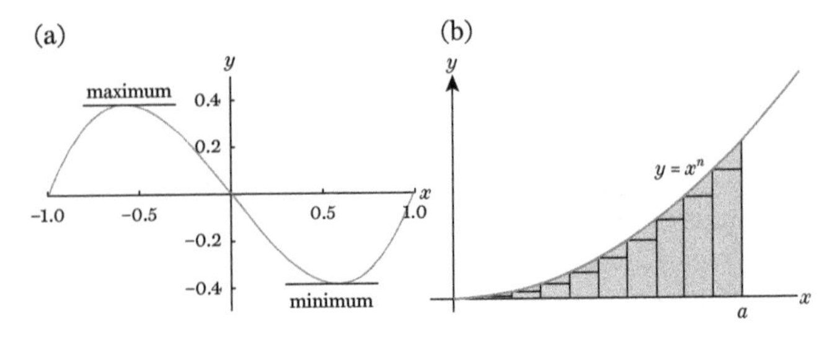

6(a). Gradients at extrema. 6(b). Area under $y = x^n$.

Fermat also considered where functions have a maximum or minimum (Figure 6(a)). At such points the gradient is zero—or equivalently the tangent line is horizontal. This result is referred to as **Fermat's theorem**. Such points are called **stationary points**.

In his *Treatise on Quadrature* of around 1658, Fermat was also able to calculate areas under the graphs of polynomials. He showed that the shaded area (Figure 6(b)) under the graph $y = x^n$, lying above the x-axis and between the lines $x = 0$ and $x = a$, equals

$$\frac{a^{n+1}}{n+1},$$

for any rational n not equal to -1. This area can be approximated by the total area of a collection of rectangles (Figure 6(b)); as the width of these rectangles becomes small, their total area gets ever closer to that under the graph. Here, again, we have a limit process being implemented, albeit in the absence of rigorous definitions at the time. Note that the above formula is nonsensical when $n = -1$, as the denominator is zero; we shall address this special case in Chapter 3 when we discuss logarithms.

Many of the key ideas of calculus appear in Fermat's work, which was fundamental to later developments. But there are important absences as well, so he is not usually credited as a founder of calculus alongside Newton and Leibniz. Fermat was working with algebraic functions which could be manipulated to a point where it was not necessary to take a limit. More than a century would pass before a rigorous notion of limit was understood.

Fermat was content to apply his methods to specific problems. He never gave a name to the gradient function of the graph of $y = f(x)$ (as $2x$ is to x^2), which would now be referred to as its *derivative* and denoted by $f'(x)$. The process of determining the derivative is called *differentiation* and the process of determining the area under a graph is called *integration*; the 'fundamental theorem' connecting these processes would need to wait until the next generation of mathematicians.

The fundamental theorem of calculus

Given a function $f(x)$, with a defined gradient $f'(x)$ everywhere, $f'(x)$ is itself a function called its **derivative** and $f(x)$ is said to be **differentiable**. Visually, we can think of $f'(x)$ as the gradient of the graph $y = f(x)$ at the point $(x, f(x))$, but it will also be useful to think of $f'(x)$ as a measure of how quickly $f(x)$ is varying as x increases. For example, if $f'(x)$ is positive then $f(x)$ increases with x and if $f'(x)$ is negative then $f(x)$ is decreasing. The process associating $f'(x)$ with $f(x)$ is called **differentiation**. Any function $F(x)$ such that $F'(x) = f(x)$ is called an **antiderivative** of $f(x)$.

Firstly, note that not all functions are differentiable. Consider the graph of the function

$$f(x) = |x| = \begin{cases} x & \text{if } x \geqslant 0, \\ -x & \text{if } x < 0, \end{cases}$$

called the **modulus function** (Figure 7). This function has a well-defined gradient of 1 for $x > 0$ and -1 for $x < 0$, but doesn't have a defined gradient at $x = 0$. Informally, this is because the graph has a corner at $x = 0$ or because the function changes

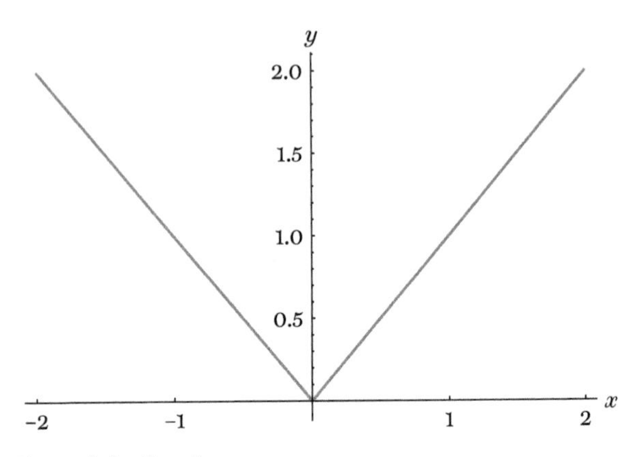

7. The modulus function.

'jerkily' there. More formally, the gradient of the chord between $(0, 0)$ and $(h, |h|)$ equals

$$\frac{f(0+h)-f(0)}{h} = \frac{|h|}{h} = \begin{cases} 1 & \text{if } h > 0, \\ -1 & \text{if } h < 0, \end{cases}$$

so that, as h becomes small, we do not get a single value for the derivative. There are yet more pathological functions that, despite being continuous, do not have a well-defined gradient at *any* point.

In this terminology, Fermat had shown that the derivative of $f(x) = x^n$ is $f'(x) = nx^{n-1}$, so that *an* antiderivative of $f(x) = x^n$ is $F(x) = \frac{x^{n+1}}{n+1}$. I write 'an' antiderivative of $f(x)$ because, for any constant c, the function $F(x) + c$ is also an antiderivative of $f(x)$. This is because adding a constant c just moves the graph of $F(x)$ up or down and so doesn't alter the gradient.

You may note that the expression for $F(x)$ looks remarkably like the formula $\frac{a^{n+1}}{n+1}$, which equals the area under the graph of $f(x)$ lying between $x = 0$ and $x = a$ (Figure 6(b)). This is a first instance of the **fundamental theorem of calculus**, which states that if $F(x)$

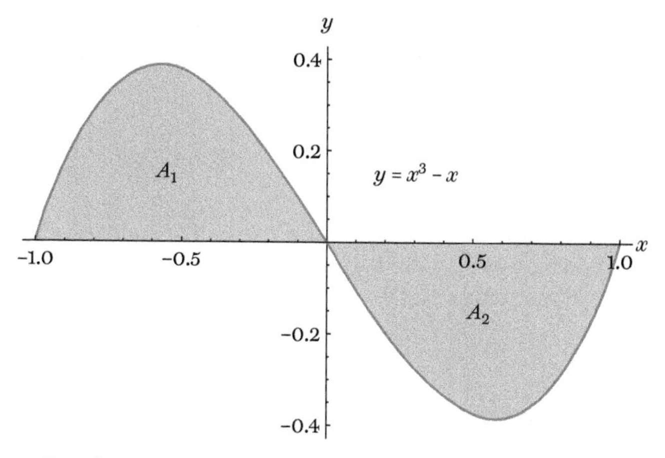

8. Signed area.

is an antiderivative of $f(x)$, then the area under the graph of $y = f(x)$, above the x-axis and between $x = a$ and $x = b$, equals

$$F(b) - F(a).$$

Note that we still arrive at the same answer if we instead use the antiderivative $F(x) + c$, as the two c terms cancel out.

In fact, 'signed area' would be a better description of what $F(b) - F(a)$ represents. Area is always positive, but when the graph of $f(x)$ is below the x-axis, the area between the graph and the x-axis contributes negatively to $F(b) - F(a)$. For example, we see above (Figure 8) a graph of $y = f(x) = x^3 - x$ which has antiderivative

$$F(x) = \frac{x^4}{4} - \frac{x^2}{2},$$

so that $F(1) - F(-1) = 0$; this does not represent the shaded area (Figure 8). Rather it represents the signed area $A_1 - A_2$, as the area A_2 is below the x-axis and so counts negatively to the total signed area; as $A_1 = A_2$, the total signed area is zero. To calculate the shaded area as a genuine area, we instead need to determine $A_1 + A_2$. Now

$$A_1 = F(0) - F(-1) = \frac{1}{4}, \qquad A_2 = -(F(1) - F(0)) = \frac{1}{4},$$

so that the total area $A_1 + A_2$ equals $\frac{1}{4} + \frac{1}{4} = \frac{1}{2}$. The standard mathematical notation for the signed area under the graph $y = f(x)$ and within the interval $a \leqslant x \leqslant b$ is

$$\int_a^b f(x)\ \mathrm{d}x,$$

so that the fundamental theorem of calculus states that

$$\int_a^b f(x)\ \mathrm{d}x = F(b) - F(a),$$

where $F(x)$ is an antiderivative of $f(x)$. The symbol $\int$ is called an *integral* sign and is an elongated 's', standing for 'sum' (or '*summa*' in Latin), and the expression on the left-hand side would be referred to as an **integral**. More precisely, it is referred to as a **definite integral**, as it has limits a and b. An **indefinite integral** is synonymous with an antiderivative.

The connection between integration and area is important to the applications of calculus, but the fundamental theorem is most easily understood in terms of rates of change. Here $\mathrm{d}x$ represents an infinitesimal increase in x so that $f(x)\mathrm{d}x = F'(x)\mathrm{d}x$ is how much $F(x)$ has increased (or decreased) during the same interval. The integral $\int_a^b F'(x)\ \mathrm{d}x$ is the sum of all these infinitesimal changes in $F(x)$ and so equals the total change in $F(x)$ as x varies from a to b, which is just $F(b) - F(a)$.

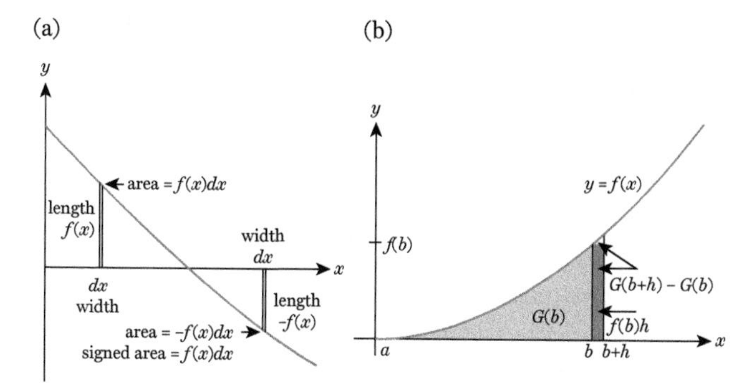

9(a). Infinitesimal rectangles. 9(b). Fundamental theorem of calculus.

In terms of area, $f(x)\mathrm{d}x$ is the area of an infinitesimal rectangle of width $\mathrm{d}x$ and height $f(x)$ which lies under the graph of $f(x)$ (Figure 9(a)). At least, this is the case if $f(x)$ is positive; if $f(x)$ is negative, then $f(x)\mathrm{d}x$ is the signed area of an infinitesimal rectangle lying under the x-axis and above the graph. Here we see a *very* informal sketch of why the fundamental theorem is true (Figure 9(b)).

Consider the signed area $G(b)$ under the graph of $f(x)$ between $x = a$ and $x = b$, thinking of a as fixed but b as varying. In the above notation

$$G(b) = \int_a^b f(x)\ \mathrm{d}x.$$

If we increase b by a very small amount h, then the area increases by exactly $G(b+h) - G(b)$. But this area is also approximately $f(b)h$, the area of the grey rectangle (Figure 9(b)). So

$$G(b+h) - G(b) \approx f(b)h,$$

where the symbol $\approx$ denotes 'approximately equals'; the two sides of the above 'equation' only differ by the area of the small white triangle above the rectangle, which is relatively negligible. Hence

$$\frac{G(b+h)-G(b)}{h}\approx f(b),$$

and as we decrease h to 0 we find in the limit that $G'(b)=f(b)$. This means $G(x)$ is an antiderivative of $f(x)$. Now $G(a)=0$ by definition—there is zero area between $x=a$ and $x=a$—so finally

$$\int_a^b f(x)\,\mathrm{d}x = G(b)-G(a)=F(b)-F(a),$$

where $F(x)$ is *any* antiderivative of $f(x)$; as commented earlier, this difference is the same for any choice of antiderivative.

A first rudimentary version of the fundamental theorem of calculus was proven using geometric methods by the Scottish mathematician James Gregory in 1668. In fact, in his unpublished work, Gregory had developed many crucial ideas of Newton and Leibniz, but his tragically premature death, aged 36, means that he is not widely remembered for his contributions. His work would not become more generally known until a memorial volume marked the tercentenary of his birth.

Significant though Fermat's contributions were, it's clearer now what was missing from his work. He never gave a name to the process of differentiation nor appreciated the inverse nature of differentiation and integration. Following Fermat, there was a coherence of understanding among mathematicians of these processes; differentiation was shown to have nice algebraic properties (see Appendix for more details); important improvements in notation were introduced; there were wide applications of calculus. But we will also see that 18th-century calculus still lacked rigour in many important aspects.

The calculus of Newton and Leibniz

Isaac Newton was one of the greatest and most influential mathematicians and scientists throughout history, and a significant figure during the Enlightenment. Besides being remembered, alongside Leibniz, as one of the developers of calculus, he made seminal contributions to the study of classical mechanics, gravity, and optics. Much of his work involved the application of calculus to real-world problems—this is evident in his three laws of motion and his law of gravitation. Gottfried Leibniz, by contrast, was a mathematician and philosopher, and was interested in producing a coherent treatment of the new calculus with well-chosen and suggestive notation, as well as the mathematical results themselves.

The previously used notation $f'(x)$ for the derivative was actually introduced by Joseph-Louis Lagrange in 1797. Leibniz began using his notation for integrals and derivatives in 1675. We have already seen his notation for the integral

$$\int_a^b f(x)\,\mathrm{d}x.$$

Leibniz envisaged $\mathrm{d}x$ as an infinitesimal increment in x so that the above is a sum of signed areas of infinitesimal rectangles. In a similar manner, if we write

$$\Delta y = f(x+h) - f(x), \qquad \Delta x = h$$

for the changes in y and x (Figure 5(b)), then $\Delta y/\Delta x$ is the gradient of PQ, which, in the limit as Q approaches P, Leibniz wrote as

$$\frac{\mathrm{d}y}{\mathrm{d}x}.$$

This is read as 'd y by d x'. Note that $\mathrm{d}y/\mathrm{d}x$ shouldn't be considered as a fraction—it is a measure of how y changes as x changes, and has no meaning as a fraction of quantities '$\mathrm{d}y$' and '$\mathrm{d}x$'.

Newton's contrasting approach shows his roots as an applied mathematician. He considered a point $P = \big(x(t), y(t)\big)$ varying on a curve with time t. He denoted the horizontal and vertical velocities of P as $\dot{x}$ and $\dot{y}$ so that

$$\dot{x} = \frac{\mathrm{d}x}{\mathrm{d}t} \qquad \text{and} \qquad \dot{y} = \frac{\mathrm{d}y}{\mathrm{d}t}$$

in Leibniz's notation, and then instead defined

$$\frac{\mathrm{d}y}{\mathrm{d}x} = \frac{\dot{y}}{\dot{x}}$$

referring to $\dot{x}$ and $\dot{y}$ as *fluxions*.

By modern standards, none of the above is satisfactorily rigorous. Leibniz is still using infinitesimals and referring to the 'infinitely small' and Newton is using undefined 'fluxions'. Neither had a rigorous sense of what a limit means, and if $\mathrm{d}x$ is to be understood as the limit of Δx as it approaches zero, then $\mathrm{d}x$ is indistinguishable from zero and $\mathrm{d}y/\mathrm{d}x$ is indistinguishable from $0/0$, which is meaningless.

Thus—and despite the many applications of calculus—there were critics of calculus' logical foundations, and these issues would remain into the 19th century. The most trenchant criticism would come from Bishop Berkeley in his 1734 work *The Analyst*. He clearly felt adherents of the calculus were trying to have their cake and eat it, writing of infinitesimals:

> They are neither finite quantities, nor quantities infinitely small, nor yet nothing.
>
> May we not call them the ghosts of departed quantities?

The quote that calculus was 'a collection of ingenious fallacies' has been ascribed to the French mathematician Michel Rolle. Whether or not Rolle actually said this, it certainly captures his concerns for the foundations of calculus.

Berkeley's and Rolle's criticisms were important ones, not least because of the wide success of the applications of calculus. Raising such questions did not detract from the impact of the work of Newton, Leibniz, and others, but did highlight the need for mathematics to get its house in order.

Newton's physics

Much of Newton's interest in calculus was due to its applications to current scientific problems. Calculus naturally finds applications in physics, as some derivatives have physical relevance. If a particle moves along the real number line so that at time t its position is $x(t)$, then the derivative

$$\frac{\mathrm{d}x}{\mathrm{d}t} \quad \text{or} \quad x'(t)$$

is its *velocity*. Note this velocity may be positive—when the particle moves left to right—or may be negative—when the particle moves right to left—or zero when the particle is stationary; by contrast, *speed* is the magnitude of velocity and cannot be negative.

The derivative of velocity is *acceleration*, which is denoted by

$$\frac{\mathrm{d}^2x}{\mathrm{d}t^2} \quad \text{or} \quad x''(t).$$

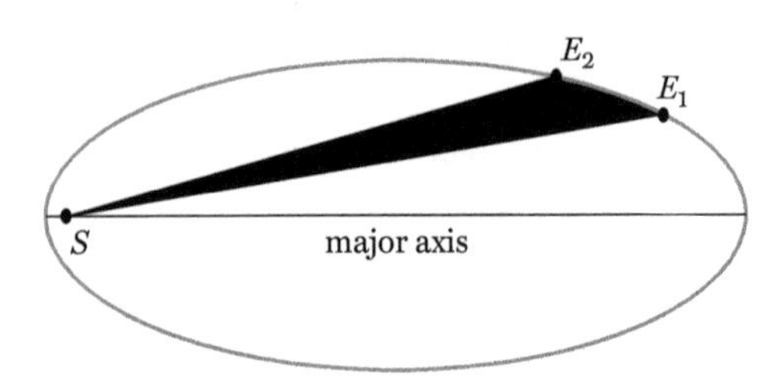

10. Kepler's first two laws.

Acceleration can be positive, zero, or negative; if it is always zero, then the particle has constant velocity; if x is increasing, then acceleration is positive/negative when the particle is speeding up/slowing down. Acceleration has importance in modelling physical systems because of *Newton's second law of motion*, which states that the force acting on a particle is equal to the particle's mass multiplied by its acceleration. Many physical applications begin with the second law; its application leads to information about a second derivative in terms of a system's current status, giving one or more *differential equations* (Chapter 3).

Between 1609 and 1619 Johannes Kepler stated three laws of planetary motion—these were laws Kepler had produced based on *observed* astronomical data. The laws, two of which are depicted in Figure 10, state:

- A planet E orbits the Sun S in an ellipse with the sun at one focus of the ellipse.
- Over a given time interval, the shaded area swept out by the line connecting the Sun and planet is always the same.
- The square of a planet's year is proportional to the cube of the orbit's major axis. The constant of proportionality is the same for each planet in the solar system.

Using his second law of motion and his *law of gravity*—that the force between two particles is proportional to each mass and inversely proportional to the square of the distance between them—Newton could mathematically *prove* all of Kepler's three laws. The details of deriving Kepler's laws mathematically are beyond this text, but applying Newton's law of gravity leads to a differential equation; solving this differential equation gives a function describing the orbit of E, which can be recognized as an ellipse. Then both the year of E's orbit and the shaded area above (Figure 10) can be represented by integrals involving the orbit's function.

The Newton–Leibniz controversy

Newton's initial work on calculus dates to 1664–6, but he would not publish on fluxions until 1704 in an appendix to his *Optiks*. His *Principia* (fully, *Philosophiae Naturalis Principia Mathematica*) of 1687 is primarily concerned with his laws of motion and gravity, though the work extensively uses arguments of calculus presented in a geometrical format. Leibniz, by contrast, did much of his work later than Newton, in 1672–6, but published his articles first, in 1684 and 1686. So there remained the question of which of the two could be credited with the invention of calculus.

The issue was somewhat muddied because in 1669 Newton had shared his work *De Analysi* with a limited number of people, some of whom Leibniz visited in London in 1672 and in Paris in 1673. This, of course, raises the question of whether Leibniz had become aware of the details of Newton's work on his visits.

The two mathematicians themselves cannot be blamed for the controversy's initial development, which was started around 1700 by other parties accusing Leibniz of plagiarizing Newton, or vice versa. The situation worsened when, in 1712, the Royal Society of London published a collection of the allegations against Leibniz; the president of the Society at the time was Isaac Newton! In 1713 the Society pronounced on the dispute, unsurprisingly finding in favour of Newton.

The modern consensus is that the two mathematicians developed the calculus independently. Ultimately, the significance of the controversy was not who was deemed to have priority at the time, but the polarizing effect the controversy had between English and continental mathematicians. English mathematics effectively cut itself off from mainstream continental mathematics, pure mathematics especially, and this situation would not be wholly remedied until the start of the 20th century. Further, because Newton's methods had been largely geometric, it was on the continent that analytic methods would progress.

Chapter 3
To the limit: analysis in the 18th and 19th centuries

e and the exponential function

Briefly digressing, consider the following scenario. You have invested a sum S of money with a banker at an annual interest rate of x, so that after one year you have earned xS and have in total $(1+x)S$. You realize that if you could convince the banker to offer half the interest rate $x/2$ twice a year, you would have $\left(1+\frac{x}{2}\right)^2 S$, which is greater—this is because

$$\left(1+\frac{x}{2}\right)^2 = 1+x+\frac{x^2}{4} > 1+x.$$

Indeed, you can improve further by having an interest rate of x/n paid n times a year so that your money grows by

$$\left(1+\frac{x}{n}\right)^n,$$

which is yet bigger and keeps getting bigger as n increases. It turns out, however gullible your banker is, that the above product reaches a limit as n becomes large.

The sequence grows as n increases but remains bounded. This means the sequence has a limit, though the limits are different for different choices of x. This 'continuous compounding' was first investigated by Thomas Harriot around 1620, and the limit defines the **exponential function**:

$$\exp(x) = \text{limit of} \left(1 + \frac{x}{n}\right)^n \text{ as } n \text{ becomes large.}$$

When $x = 1$ the value $\exp(1)$ is denoted as *e*. After π, *e* is the second most important constant in all of mathematics.

As a value, $e = 2.718281828459045\ldots$ and has been calculated to over 10^{13} places. The great Swiss mathematician Leonhard Euler showed in 1737 that *e* is irrational—that is, *e* is not the ratio of two integers (see Appendix for a proof)—and in 1873 Charles Hermite proved that *e* is *transcendental*—meaning that *e* is not the solution of any polynomial equation with whole-number coefficients. The notation *e* was introduced by Euler and *e* is often referred to as *Euler's number*, because of his clarifying work on the exponential function, despite the earlier studies of Harriot (and also Jacob Bernoulli).

A graph of the exponential function can be seen below (Figure 11(a)).

The phrase 'exponential' is figuratively used to describe rapid growth; in a technical sense growth is exponential if, over a given period, the same relative growth occurs. The same is true of

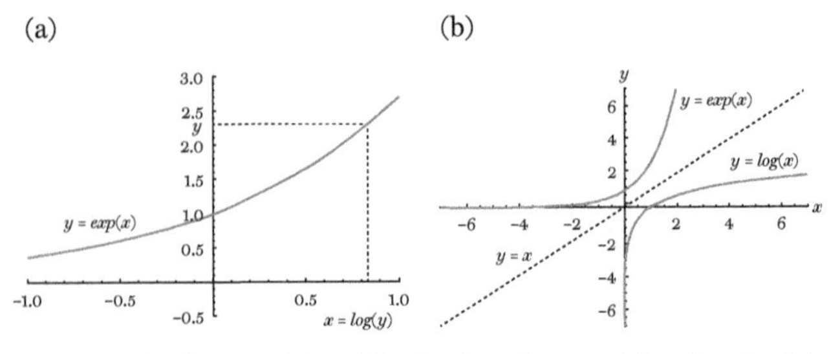

11(a). Graph of $y = \exp(x)$. 11(b). Graphs of $y = \exp(x)$ and $y = \log(x)$.

exponential decay; for example, the *half-life* of a radioactive element is the time taken for half the atoms to decay into other elements, and this time is constant irrespective of the amount of material present. Consequently, the exponential function satisfies the identity

$$\exp(x+y) = \exp(x)\exp(y)$$

for any real numbers x and y. Note that this means for a counting number n that

$$\begin{aligned} \exp(n) &= \exp(n-1)\exp(1) \\ &= \exp(n-2)\exp(1)^2 \\ &= \quad \vdots \\ &= \exp(1)^n \\ &= e^n. \end{aligned}$$

However, at this stage, we cannot simply write $\exp(x) = e^x$ for a general real number x. The notion that e^x means 'e times by itself x times' is nonsensical when, say, $x = \sqrt{2}$. But we will see we can use the exponential function to *define* general powers.

Logarithms and powers

It makes sense to define a^n as 'a multiplied by itself n times' only when n is a positive whole number. When a is positive, we can make sense of rational powers such as $a^{2/3}$ as the cube root of a^2, but it remains unclear how a general power a^x might be defined. Though it seems reasonable to still expect

$$a^{x+y} = a^x a^y$$

for any real numbers x and y. In order to define arbitrary powers, we need to introduce *logarithms*.

From the graph in Figure 11(a), we see the exponential function attains all positive values y. Moreover, as $\exp(x)$ is increasing, there is a *unique* value x such that

$$\exp(x) = y.$$

This value x is the **logarithm** (or *natural logarithm*) of y, and we write $x = \log(y)$, a common alternative notation being $x = \ln(y)$. Importantly, the logarithm function has the property

$$\log(xy) = \log(x) + \log(y)$$

for positive x, y; this identity corresponds to the previous identity for $\exp(x + y)$ (see Appendix).

For positive a, we can now *define* general powers in terms of this logarithm function as

$$a^x = \exp(x \log(a)).$$

This gives the desired earlier algebraic property as

$$\begin{aligned} a^x a^y &= \exp(x \log(a)) \exp(y \log(a)) \\ &= \exp(x \log(a) + y \log(a)) \\ &= \exp((x + y)\log(a)) = a^{x+y}. \end{aligned}$$

Note that a^x agrees with our previous notion of a^x equalling a multiplied by itself x times when x is a counting number.

When $a = e$, so that $\log(e) = 1$, then $e^x = \exp(x)$, and so we will write e^x rather than $\exp(x)$ from now on. The function a^x defines a differentiable function of x which has derivative $a^x\log(a)$ (see Appendix). So when $a = e$, then e^x equals its own derivative—an important property of the exponential function and one we will return to. The derivative of $\log(x)$ equals $\frac{1}{x}$. Note this is the antiderivative we have been missing for the powers x^n. The antiderivative of x^n is $\frac{x^{n+1}}{n+1}$ when n does not equal -1, and in the case $n = -1$ the antiderivative is $\log(x)$.

Before modern calculators and computers, logarithms were an important part of how products were determined because sums are much easier to calculate than products. Logarithms were first introduced in 1614 by the Scottish mathematician John Napier. His somewhat different definition for $\mathrm{Nlog}(x)$, the Napierian logarithm of positive x, is the value of y satisfying

$$x = 10^7\left(1 - \frac{1}{10^7}\right)^y,$$

which satisfies the identity

$$\mathrm{Nlog}\left(\frac{x_1 x_2}{10^7}\right) = \mathrm{Nlog}(x_1) + \mathrm{Nlog}(x_2).$$

Between 1617 and 1624 Napier's tables were improved by Henry Briggs, who introduced *common logarithms* or logarithms to base 10, better suited to the decimal system. The common logarithm y for positive x satisfies $x = 10^y$ and is written $\log_{10} x$.

Tables of logarithms reduced difficult products to simpler sums. For example, suppose we wished to calculate

$$230.1367 \times 1213.9743.$$

A table of common logarithms need only contain the logarithms for numbers between 1 and 10 as

$$\log_{10} 230.1367 = \log_{10}(10^2 \times 2.301367) = 2 + \log_{10} 2.301367,$$
$$\log_{10} 1213.9743 = \log_{10}(10^3 \times 1.2139743) = 3 + \log_{10} 1.2139743.$$

On looking up these logarithms, we would find

$$\log_{10} 2.301367 = 0.361985881..., \log_{10} 1.2139743 = 0.084209483....$$

(It may be that the tables only provide logarithms for inputs given to fewer decimal places, but then it is possible to *interpolate* to find logarithms for values between those given in the tables.) Then $\log_{10}(230.1367 \times 1213.9743)$ equals

$$2.361985881\ldots + 3.084209483\ldots = 5.446195374\ldots,$$

an addition which is simple compared with the earlier multiplication. Returning to our tables, we would find that 0.446195374 is the common logarithm of 2.793800392, and hence

$$230.1367 \times 1213.9743 = 10^5 \times 2.793800392 = 279380.0392$$

to four decimal places. The precise answer is found quickly with a modern calculator to be 279380.03928681, but it is easy to forget how recent an invention electronic calculators and computers are. Logarithm tables (and *slide rules*, which make use of logarithmic scales) were widely used until the early 1970s.

Power series and Taylor series

Power series are infinite sums of the form

$$f(x) = c_0 + c_1 x + c_2 x^2 + c_3 x^3 + \cdots.$$

They provide a very powerful tool in analysis, as demonstrated by Newton in his *De Analysi* and later by Euler and Lagrange.

The real numbers $c_0, c_1, c_2, \ldots$ are considered fixed (for this power series), whilst the real number x is considered an input which may vary. Depending on the value of x, the above sum may converge or not. When $x = 0$ we find that $f(0) = c_0$, and so the sum definitely converges. But for $x = 1$, say, we see

$$f(1) = c_0 + c_1 + c_2 + \cdots,$$

which may or may not converge. As an example, consider the power series

$$f(x) = 1 + x + x^2 + x^3 + x^4 + \cdots,$$

where $c_n = 1$ for each n. When $x = 1$ we can see that $f(1) = 1 + 1 + 1 + \cdots$ doesn't converge. When $x = -1$ we obtain Grandi's sum $f(-1) = 1 - 1 + 1 - 1 + \cdots$, which also doesn't converge. In fact, it can be shown that $f(x)$ converges precisely when $-1 < x < 1$. For such x we can validly argue by the algebra of limits that:

$$f(x) = 1 + x + x^2 + x^3 + \cdots = 1 + x(1 + x + x^2 + \cdots) = 1 + xf(x),$$

which rearranges to give

$$f(x) = \frac{1}{1-x}.$$

Hence

$$f(x) = 1 + x + x^2 + \cdots = \begin{cases} 1/(1-x) & \text{if } -1 < x < 1, \\ \text{undefined} & \text{for other } x. \end{cases}$$

Note the function $1/(1-x)$ is defined more generally, namely whenever x does not equal 1, and so should be considered a different function to $f(x)$. Rather, $f(x)$ is a power series representation of $1/(1-x)$, locally defined just on the interval $-1 < x < 1$.

A general power series defines a function $f(x)$ for those x where the power series converges. This will occur for x in an interval $-R < x < R$, for some R in the range $0 \leqslant R \leqslant \infty$. R is called the *radius of convergence*, and we set $R = \infty$ when the series converges for all values of x. The power series does not converge when $x > R$ or $x < -R$, and may or may not converge when $x = R$ or $x = -R$. In the given example, $R = 1$. Importantly, a power series defines a differentiable function $f(x)$ where it converges and the derivative $f'(x)$ can be found by differentiating the power series term-by-term. Recalling that the derivative of x^n equals nx^{n-1}, we obtain

$$f'(x) = c_1 + 2c_2x + 3c_3x^2 + 4c_4x^3 + \cdots,$$

and this can be repeatedly applied to find power series for the higher derivatives of $f(x)$. Conversely, we might ask: what functions can be represented by a power series? Certainly, such a function must be repeatedly differentiable. We might begin with a function $f(x)$ and seek to find a power series representing it so that

$$f(x) = c_0 + c_1x + c_2x^2 + c_3x^3 + \cdots.$$

Our problem is to determine $c_0, c_1, c_2, \ldots$ We can quickly find c_0 by setting $x = 0$ in the above so that $c_0 = f(0)$. And if we repeatedly differentiate the above and set $x = 0$, we find

$$\begin{aligned}
f'(x) &= c_1 + 2c_2x + 3c_3x^2 + 4c_4x^3 + \cdots &\text{giving}\quad c_1 &= f'(0),\\
f''(x) &= 2c_2 + 6c_3x + 12c_4x^2 + 20c_5x^3 + \cdots &\text{giving}\quad 2c_2 &= f''(0),\\
f'''(x) &= 6c_3 + 24c_4x + 60c_5x^2 + 120c_6x^3 + \cdots &\text{giving}\quad 6c_3 &= f'''(0).
\end{aligned}$$

So

$$c_1 = f'(0), \qquad c_2 = \frac{f''(0)}{1 \times 2}, \qquad c_3 = \frac{f'''(0)}{1 \times 2 \times 3},$$

and, generally, we find that

$$c_n = \frac{f^{(n)}(0)}{n!},$$

where $f^{(n)}(x)$ denotes the nth derivative of $f(x)$, the function arrived at when $f(x)$ is differentiated n times, and where

$$n! = 1 \times 2 \times 3 \times \cdots \times n,$$

which is read as 'n **factorial**'. So, we might expect that

$$f(x) = f(0) + \frac{f'(0)}{1!}x + \frac{f''(0)}{2!}x^2 + \frac{f'''(0)}{3!}x^3 + \cdots + \frac{f^{(n)}(0)}{n!}x^n + \cdots.$$

This is called the **Taylor series** of $f(x)$ centred at 0 (also known as the *Maclaurin series* of $f(x)$), and the same reasoning can be applied to arrive at the general Taylor series centred at a real number a:

$$f(x) = f(a) + \frac{f'(a)}{1!}(x-a) + \frac{f''(a)}{2!}(x-a)^2 + \frac{f'''(a)}{3!}(x-a)^3 + \cdots + \frac{f^{(n)}(a)}{n!}(x-a)^n + \cdots.$$

These series are named after Brook Taylor, who first published on them in 1715, and Colin Maclaurin, though Newton had been aware of the general form of Taylor series as early as 1691, and James Gregory possibly even earlier.

Recall we noted earlier that the derivative of e^x is e^x. In fact, the exponential function $f(x) = e^x$ can be uniquely characterized by the properties

$$f'(x) = f(x) \qquad \text{and} \qquad f(0) = 1.$$

This means that $f^{(n)}(x) = e^x$ and $f^{(n)}(0) = 1$ for all n; hence the Taylor series of the exponential function is

$$e^x = 1 + x + \frac{x^2}{2!} + \frac{x^3}{3!} + \cdots + \frac{x^n}{n!} + \cdots.$$

This series converges for all real x. It also gives us a new definition for $e = e^1$, namely

$$e = 1 + 1 + \frac{1}{2!} + \frac{1}{3!} + \cdots + \frac{1}{n!} + \cdots,$$

which converges to e much faster than the previous limit of Harriot and Bernoulli.

Unfortunately, the problem of which real functions can be represented by power series is not particularly simple. Given a repeatedly differentiable function $f(x)$, it is possible to write down its Taylor series centred at 0 as above. That Taylor series necessarily agrees with $f(x)$ at $x = 0$, but there are examples where the Taylor series and the function agree *only* at $x = 0$. A function is said to be **analytic** at a point if it is repeatedly differentiable *and* agrees with its Taylor series on an interval around that point. We will see (Chapter 7) that this situation resolves much more simply with complex functions.

Radians and trigonometry

Figure 12(a) shows a triangle ABC with a right angle at C. Denoting the angle $\angle BAC$ as x, we can define two functions **sine** and **cosine** of x, denoted by $\sin(x)$ and $\cos(x)$, as

$$\sin(x) = \frac{\text{length of } BC}{\text{length of } AB}, \qquad \cos(x) = \frac{\text{length of } AC}{\text{length of } AB}.$$

These are functions of the angle x, rather than the triangle ABC: if we scale the triangle to $AB'C'$, the three sides scale by the same factor and so the above fractions stay the same. The functions sine and cosine are referred to as *trigonometric ratios* or *trigonometric functions*. Pythagoras' theorem states that

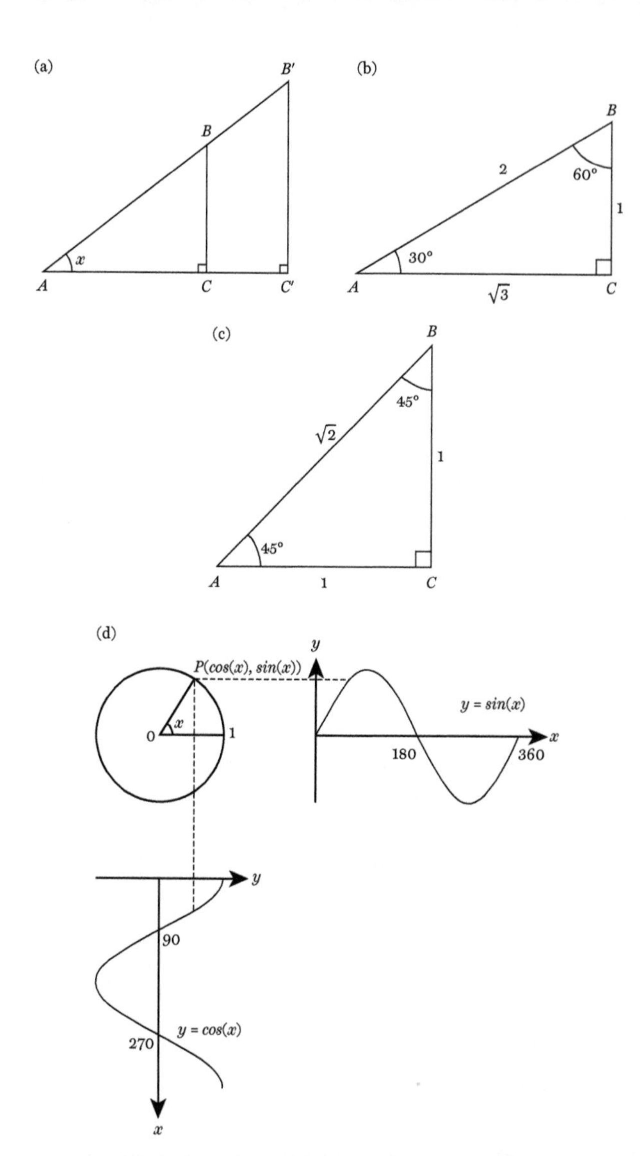

12(a). A right-angled triangle. 12(b). $x = 30^\circ$ and $x = 60^\circ$. 12(c). $x = 45^\circ$. 12(d). The circular functions' graphs.

$$(\text{length of } AB)^2 = (\text{length of } AC)^2 + (\text{length of } BC)^2,$$

and so the following identity holds:

$$\left(\cos(x)\right)^2 + \left(\sin(x)\right)^2 = \left(\frac{\text{length of } AC}{\text{length of } AB}\right)^2 + \left(\frac{\text{length of } BC}{\text{length of } AB}\right)^2 = 1.$$

This geometric argument works for angles x up to a right angle.

Sine and cosine are also called the *circular functions*, as a general point P on the circle centred on the origin O and with radius 1 has co-ordinates $(\cos(x), \sin(x))$ for some x (Figure 12(d)). Further, as the point P traces out the whole circle, we can draw out the graphs of $\sin(x)$ and $\cos(x)$. It is then apparent that sine and cosine have a period of 360°. That is,

$$\sin(x + 360) = \sin(x), \qquad \cos(x + 360) = \cos(x),$$

because after P has moved on 360°, a whole revolution, it has returned to the same point of the circle.

When angles are first introduced, they are typically measured in *degrees*, which has the symbol °. There are 360° in a whole angle, and 90° in a right angle. From the definition, and from considering the drawn triangles (Figures 12(b), 12(c)), we can see that

$$\sin(0^\circ) = \cos(90^\circ) = 0, \quad \sin(30^\circ) = \cos(60^\circ) = \frac{1}{2},$$
$$\sin(45^\circ) = \cos(45^\circ) = \frac{1}{\sqrt{2}},$$
$$\sin(60^\circ) = \cos(30^\circ) = \frac{\sqrt{3}}{2}, \quad \sin(90^\circ) = \cos(0^\circ) = 1.$$

However, mathematicians, with good reason, use **radians** to measure angles; the benefits of using radians are particularly clear in calculus. There are 2π radians in a whole angle, rather than 360°, so that 1 radian equals $\frac{360^\circ}{2\pi}$, roughly 57°. The corresponding values for sine and cosine, when using radians, are given below:

$$\sin(0)=0,\quad \sin\left(\frac{\pi}{6}\right)=\frac{1}{2},\quad \sin\left(\frac{\pi}{4}\right)=\frac{1}{\sqrt{2}},\quad \sin\left(\frac{\pi}{3}\right)=\frac{\sqrt{3}}{2},\quad \sin\left(\frac{\pi}{2}\right)=1,$$
$$\cos(0)=1,\quad \cos\left(\frac{\pi}{6}\right)=\frac{\sqrt{3}}{2},\quad \cos\left(\frac{\pi}{4}\right)=\frac{1}{\sqrt{2}},\quad \cos\left(\frac{\pi}{3}\right)=\frac{1}{2},\quad \cos\left(\frac{\pi}{2}\right)=0.$$

In geometry, some formulae are improved with radians—for example, the length of a circular arc, radius r and angle x, equals rx rather than the messier formula seen when using degrees, $\frac{\pi r x}{180}$. In calculus, though, the use of radians is crucial.

Firstly, the derivative of $\sin(x)$ is $\cos(x)$ and the derivative of $\cos(x)$ is $-\sin(x)$, *but these facts are true only if we are using radians to measure angles*. Secondly, from these derivatives we can determine the Taylor series for sine and cosine. The successive derivatives of $\sin(x)$ are

$$\sin(x),\quad \cos(x),\quad -\sin(x),\quad -\cos(x),\quad \sin(x),\ \ldots,$$

repeating every four. Setting $x=0$, we obtain the sequence $0, 1, 0, -1, 0, 1, 0, -1$, so the Taylor series for sine is given by

$$\begin{aligned}\sin(x) &= 0+\frac{1}{1!}x+\frac{0}{2!}x^2+\frac{-1}{3!}x^3+\frac{0}{4!}x^4+\frac{1}{5!}x^5+\cdots\\ &= x-\frac{x^3}{3!}+\frac{x^5}{5!}-\frac{x^7}{7!}+\cdots.\end{aligned}$$

Likewise the cosine series can be determined as

$$\begin{aligned}\cos(x) &= 1+\frac{0}{1!}x+\frac{-1}{2!}x^2+\frac{0}{3!}x^3+\frac{1}{4!}x^4+\frac{0}{5!}x^5+\cdots\\ &= 1-\frac{x^2}{2!}+\frac{x^4}{4!}-\frac{x^6}{6!}+\cdots.\end{aligned}$$

These series converge for all real x and *again are correct, provided radians are used.* They were first determined in Europe by Newton and appear in his *De Analysi*, but had been known to Madhava centuries earlier. Analytically, it is beneficial to take them as our definitions for sine and cosine. Term-by-term differentiation can be validly applied to these series and, further, the rules of differentiation can be used to prove identities such as $(\sin(x))^2 + (\cos(x))^2 = 1$ for all real x. (See Appendix. Also appearing in the Appendix is a derivation of Madhava's series for π using Taylor series.)

Euler

Leonhard Euler (Figure 13(a)), pronounced 'oil-er', was a prolific Swiss mathematician and a titan of 18th-century mathematics, with over 800 papers bearing his name. He made major contributions across mathematics—number theory, fluid dynamics, calculus of variations (Chapter 5), complex functions (Chapter 7)—but especially in the study of infinite sums. He is particularly remembered for evaluating the infinite sum S where

$$S = 1 + \frac{1}{2^2} + \frac{1}{3^2} + \frac{1}{4^2} + \frac{1}{5^2} + \cdots = \frac{\pi^2}{6},$$

the so-called *Basel problem*. He also produced some of the first *topological* results, showing that it's impossible to traverse the seven bridges in Königsberg without repetition (Figure 13(b)), and he showed that for a (convex) polyhedron $V - E + F = 2$, where V, E, F respectively denote the number of vertices (corners), edges, and faces of the polyhedron. Both these results depend on shape (e.g. how the points are connected) rather than geometry (e.g. the lengths of the edges). Much modern mathematical notation is due to him; he introduced the notation e for Euler's number, i for the square root of -1 (Chapter 7), and f for a function, and popularized the notation for π.

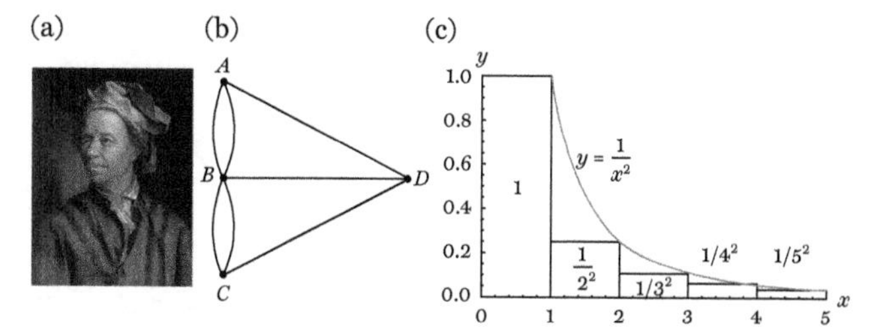

13(a). Leonhard Euler. 13(b). Königsberg's bridges as a graph. 13(c). Showing S converges.

We can use calculus to show that the sum S is finite. Note that the rectangles under the graph of $y = 1/x^2$ (Figure 13(c)) have total area S which is then less than

$$1 + \int_1^\infty \frac{1}{x^2}\,\mathrm{d}x,$$

as this equals the area of the first rectangle and the area under the graph for $x \geqslant 1$. As $-1/x$ is an antiderivative of $1/x^2$, the fundamental theorem shows

$$1 + \int_1^X \frac{\mathrm{d}x}{x^2} = 1 + \left(\frac{-1}{X} - \frac{-1}{1}\right) = 2 - \frac{1}{X},$$

which approaches 2 as X becomes large. It follows that S is finite and less than 2.

By modern standards, Euler's first solution of the Basel problem was inventive but cavalier; details are in the Appendix. This highlights how, despite all the progress and creativity of the 18th century, mathematical rigour had not progressed far. And whilst Euler, in his seminal *Introductio in Analysin Infinitorum* of 1748, made central for the first time the notion of function, his definition—that a function is described by a single analytic expression—would ultimately prove too restrictive and even caused difficulties at the time for the physical systems mathematicians were modelling (as we will note with the wave equation in Chapter 6).

Differential equations

The greatest application of calculus is a consequence of how well the world around us can be modelled by differential equations. Differential equations describe theories of gravity (Newtonian and Einsteinian), electromagnetism (Maxwell's equations, wave equation, Laplace's and Poisson's equations), classical mechanics (Lagrange's and Hamilton's equations), quantum theory (Schrödinger's equation), fluid dynamics (Navier–Stokes equations), economics and finance (Black–Scholes equation), thermodynamics (heat equation), and mathematical biology (predator–prey interactions and epidemiology), etc.

A **differential equation** is an equation involving a function and its derivatives. Such equations arise quite naturally—for example, applying Newton's second law of motion to any system states something about acceleration, a second derivative. Suppose a particle moves vertically under gravity, having height $h(t)$ over the ground at time t. Ignoring air resistance and assuming gravity (denoted as g) is constant, $h(t)$ satisfies the differential equation

$$h''(t) = -g.$$

This differential equation has **order** 2, or is *second order*, as the highest derivative involved is the second derivative.

This equation states that the particle is accelerating due to gravity, and the minus sign denotes that gravity acts downwards. We can find the general solution for $h(t)$ by integrating twice. An antiderivative of $-g$ is $-gt$, so the general antiderivative is $-gt + c_1$, where c_1 is a constant, giving

$$h'(t) = -gt + c_1.$$

And an antiderivative of $-gt + c_1$ is $-\frac{1}{2}gt^2 + c_1 t$, so that

$$h(t) = -\frac{1}{2}gt^2 + c_1 t + c_2,$$

where c_2 is a second constant. This does not specify the particle's trajectory without further information; rather the above expression describes *all* possible flights. But, say, knowing the initial height $h(0) = c_2$ and initial velocity $h'(0) = c_1$ determines $h(t)$ exactly. Such a description of a system—a differential equation and initial conditions—is called an **initial value problem.**

Exponential growth can also be characterized by a differential equation. Physically, this might represent the growth in the number $N(t)$ of bacteria with time t from a single bacterium, while there is sufficient food or energy resource. If r, a positive constant, is the growth rate of the bacteria, then $N(t)$ satisfies the initial value problem

$$N'(t) = rN(t), \qquad N(0) = 1.$$

Note $N(t)$ can be a general positive number, whilst the number of bacteria is a whole number, but the above provides a reasonable approximation of the reality of the system.

Differential equations can often be solved using power series. We might *try* a solution of the form

$$N(t) = c_0 + c_1 t + c_2 t^2 + c_3 t^3 + \cdots,$$

and substitute this into the above differential equation. Using the initial condition, we can see that $c_0 = N(0) = 1$. And differentiating term-by-term, the differential equation now reads as

$$c_1 + 2c_2t + 3c_3t^2 + 4c_4t^3 + \cdots = r + rc_1t + rc_2t^2 + rc_3t^3 + \cdots.$$

Comparing the coefficients of like powers on the left- and right-hand sides gives

$$c_1 = r, \quad c_2 = \frac{rc_1}{2} = \frac{r^2}{2!}, \quad c_3 = \frac{rc_2}{3} = \frac{r^3}{3!}, \quad c_4 = \frac{rc_3}{4} = \frac{r^4}{4!},$$

and so on. From this we can see that the solution is

$$N(t) = 1 + rt + \frac{r^2t^2}{2!} + \frac{r^3t^3}{3!} + \frac{r^4t^4}{4!} + \cdots = e^{rt}.$$

This method is valid, when it works, in that it will yield *a* solution, but not all differentiable functions can be represented by power series.

The solution $N(t) = e^{rt}$ cannot be realistic for all times t, as the function grows without bound. Eventually resources will become limited, so we might introduce a population capacity K to make the model more realistic as Pierre Verhulst did in 1838 with the differential equation

$$N'(t) = rN(t)\left(1 - \frac{N(t)}{K}\right).$$

His equation has a drag factor of $1 - N/K$ on the growth rate. When N is small compared with K, the growth rate is still approximately r, and N grows almost exponentially, but as N approaches the capacity K, the growth rate becomes close to zero. The S-like solution to Verhulst's equation appears in Figure 14.

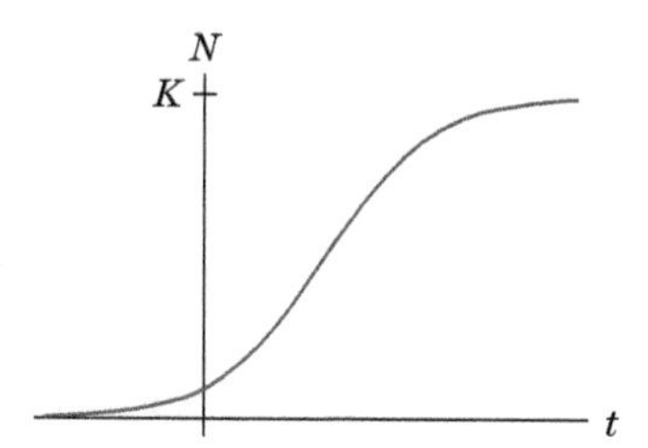

14. S-like growth for Verhulst's model.

In these examples a single variable, h or N, determines the status of a system, but some systems involve more than one variable. For example, a system involving competing populations of foxes $F(t)$ and rabbits $R(t)$ needs both population sizes to be specified. Equations modelling such systems were studied by the American biologist Alfred Lotka and Italian mathematician Vito Volterra, who independently arrived at the differential equations

$$F'(t) = -mF + aFR, \qquad R'(t) = bR - kFR,$$

where a, b, k, m are positive constants. The two equations model the system by assuming:

- the rabbits breed at a certain rate b;
- the number of rabbits being killed, $-kFR$, is proportional to both the rabbit population (the more rabbits there are, the more get caught) and the number of foxes (the more predators, the more rabbits killed);
- the foxes rely on the rabbits as a food source to multiply, so, for the reasons just given, their growth term is an FR term countered by a term, $-mF$, proportional to F, due to death from disease and old age.

These *simultaneous* differential equations lead to periodic solutions $F(t), R(t)$, which are plotted in Figure 15(a). Some of the different possible egg-shaped paths that the point $(F(t), R(t))$

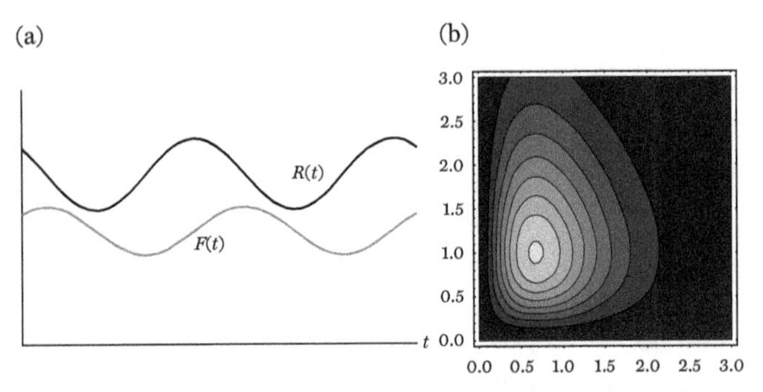

15(a). Fox and rabbit populations. 15(b). The cycles (F, R) travels.

might travel are also plotted here (Figure 15(b)). If, at a certain time, we had F_0 foxes and R_0 rabbits, the point (F_0, R_0) would lie on one such egg-shaped path; as time progresses the point $(F(t), R(t))$ would move around that path, eventually returning to the same point (F_0, R_0) and repeating forever.

Émile Picard showed, given a set of reasonable though technical criteria, that there exists a unique solution to the initial value problem

$$\frac{\mathrm{d}y}{\mathrm{d}x} = f(x, y), \qquad y(x_0) = y_0,$$

which is locally defined. Picard's proof is constructive, defining a sequence of functions $y_n(x)$ which converge to the solution. The sequence is defined iteratively by

$$y_0(x) = y_0, \qquad y_n(x) = y_0 + \int_0^x y_{n-1}(t)\ \mathrm{d}t.$$

As an example, the exponential function satisfies this initial value problem when $f(x, y) = y, \quad x_0 = 0, \quad y_0 = 1$. In this case, Picard's theorem generates the functions

$$
\begin{aligned}
y_0(x) &= 1, \\
y_1(x) &= 1 + \int_0^x 1 \, \mathrm{d}t = 1 + x, \\
y_2(x) &= 1 + \int_0^x (1+t) \, \mathrm{d}t = 1 + x + \frac{x^2}{2}, \\
y_3(x) &= 1 + \int_0^x \left(1 + t + \frac{t^2}{2}\right) \mathrm{d}t = 1 + x + \frac{x^2}{2} + \frac{x^3}{6}.
\end{aligned}
$$

We can see this sequence equals the partial sums of the exponential function's Taylor series.

It may seem hard to imagine a situation where a solution is not defined, at least locally. An initial value problem provides a point on the solution (x_0, y_0) and a 'direction of travel' by specifying the gradient there. It could become the case that following that direction of travel might lead to $y(x)$ or $y'(x)$ becoming infinite (Figures 16(a), 16(b)). So it's reasonable that the solution may only be locally defined.

But if Picard's criteria are not met, then there can be more than one way to proceed along the direction of travel (Figure 16(c)). The general solution to the initial value problem

$$
\frac{\mathrm{d}y}{\mathrm{d}x} = 2\sqrt{y}, \qquad y(0) = 0
$$

is the infinite family of functions

$$
y(x) = \begin{cases} 0 & \text{if } x \leqslant a, \\ (x-a)^2 & \text{if } x > a, \end{cases}
$$

where $0 \leqslant a \leqslant \infty$. Note that when $y = 0$, the gradient $\mathrm{d}y/\mathrm{d}x$ is also zero. The issue is that while y remains zero, there are two possible ways to follow the given direction of travel: continue along

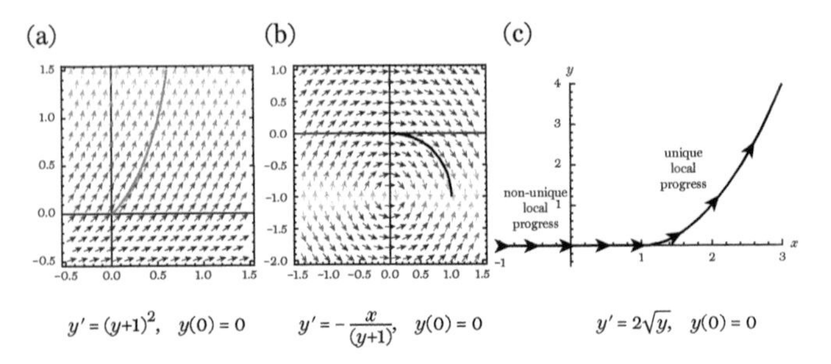

16(a). $y(x)$ becomes infinite. 16(b). $y'(x)$ becomes infinite. 16(c). Non-unique progress.

the x-axis or start on a half-parabola. But once the solution moves on to a parabola, the solution must continue along it.

Bolzano and Weierstrass

Finally, in the mid 19th century, a definition of limit would be found which made no reference to infinitesimals. This definition is usually attributed to the German mathematician Karl Weierstrass, though such definitions had been implicit in Cauchy's *Cours d'Analyse* of 1821 and explicitly in the work of 1810–17 of Bernard Bolzano, which went unnoticed in his lifetime.

We already met in Chapter 1 the definition of a sequence $x_1, x_2, x_3, \ldots$ having a **limit** L. This meant that, however close we wished the sequence to get to L, this would eventually happen and continue so. More formally, this means that given any positive ε (this is our notion of 'close', so we typically consider ε as small), there exists a positive integer N (this is a point from which 'eventually' starts happening) for which x_n is suitably close to L when $n \geqslant N$.

We similarly wish to define what it means for a real function $f(x)$ to have a limit L as x gets near to a. Note, generally, that $f(a)$ may

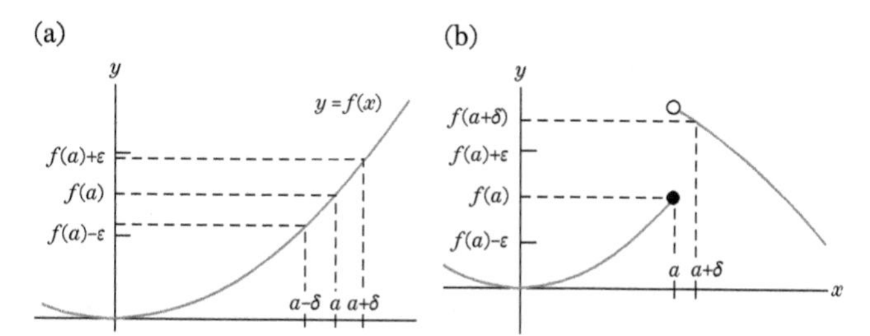

17(a). A continuous function. 17(b). A discontinuous function.

not be defined and, even if it is, $f(a)$ may be different to L. If $L = f(a)$, then $f(x)$ is said to be **continuous** at a (Figure 17(a)); likewise, the function may have no limit at a and distinct limits from the left and right (Figure 17(b)). The study of continuous real functions, and their generalization to metric spaces and topological spaces, is an important part of analysis, but one which is covered in detail in *Topology: A Very Short Introduction.*

We want to guarantee $f(x)$ is sufficiently close to L if x is sufficiently close to a. So we require, for any positive ε (the demanded closeness in the outputs), that there exists a positive δ (a closeness in the inputs to meet the demand) such that $f(x)$ is within ε of L whenever x is within δ of a. It's important to note that:

- the definition makes no reference to infinitesimals;
- we require the output $f(x)$ to be constrained in a certain way if x is appropriately constrained;
- we need to be able to do this for *all* constraints ε: for each choice of ε we will need a choice of δ that meets the requirement;
- for a smaller ε, δ will usually need to be smaller as well;
- given ε, any δ that meets the requirement is fine—we're not looking for a largest δ, say;
- the 'faster' the function $f(x)$ is changing at a, the smaller δ will need to be, relative to ε.

So, we can finally and rigorously define what it means for the function $f(x)$ to be differentiable at a. This means that the approximating gradient

$$\frac{f(a+h)-f(a)}{h}$$

has a limit as h approaches 0, and we denote this limit as $f'(a)$.

This terminology is often referred to as ε-δ analysis (read 'epsilon-delta'). This is the standard notation of analysis, and is a final shift from the informality or imprecision of previous definitions. It is the hallmark of modern analysis, which by Weierstrass' time had carved out for itself a canon of its own within mathematics. Intuitively, you may think of a continuous function as one with a graph which can be drawn without taking your pen off the paper, and that a function is differentiable if the graph does not change jerkily, but both these notions are woefully insufficient for proving results about continuous and differentiable functions.

Riemann's integral

During the 19th century, integration theory also advanced considerably. In his *Résumé* of 1823, Cauchy defined integrals of continuous functions on intervals $a \leqslant x \leqslant b$. His work is important in various ways: he considered integration in the context of signed areas, rather than as simply antidifferentiation—the inverse of differentiation—and he (largely) proved the fundamental theorem ('largely' as there was an important technical gap in his argument). However, Cauchy's treatment, of functions that were either continuous or had only finitely many discontinuities, wasn't sufficiently broad for the evolving notion of a function.

In 1854 Riemann developed a more general theory of integration, treating bounded functions on bounded intervals, though what follows below is an equivalent treatment of Riemann's integral by

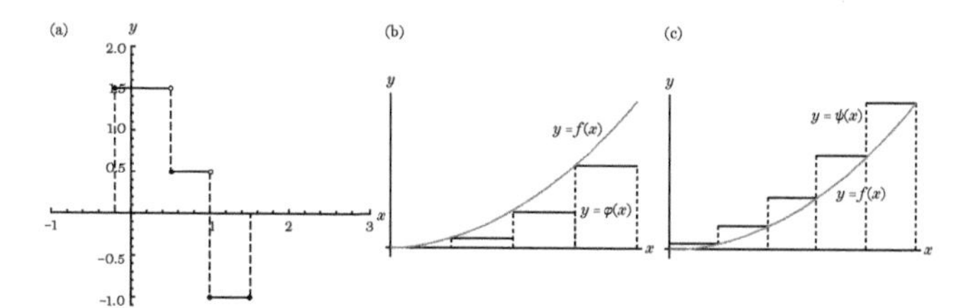

18(a). A step function. 18(b). A step function below $f(x)$.
18(c). A step function above $f(x)$.

the French mathematician Gaston Darboux from 1875. Note that, compared with 17th- and 18th-century notions of an integral, this definition makes no reference to infinitesimals.

As an uncontroversial starting point, we define the area of a rectangle as equal to its base multiplied by its height. A **step function** $\varphi(x)$, on an interval $a \leqslant x \leqslant b$, is a function whose graph comprises a finite collection of rectangles, above or below the x-axis (Figure 18(a)). The integral of a step function is just the sum of these rectangles' signed areas.

Our aim is to assign a real number I to the integral of a bounded function $f(x)$. If the graph of a step function $\varphi(x)$ lies entirely below the graph of $f(x)$, then we would expect I to be at least the integral of $\varphi(x)$ (Figure 18(b)). Likewise, if the graph of a step function $\psi(x)$ lies above the graph of $f(x)$, then we would expect I to be at most the integral of $\psi(x)$ (Figure 18(c)). For most naturally occurring functions these requirements specify a unique value for I, but we will see in Chapter 8 that this is not generally the case.

More explicitly, given a bounded function $f(x)$ on $a \leqslant x \leqslant b$, the *lower Riemann integral* of $f(x)$ is the smallest real number I_{lower} such that

$$I_{\text{lower}} \geqslant \int_a^b \varphi(x)\ \mathrm{d}x,$$

where $\varphi(x)$ is a step function with $\varphi(x) \leqslant f(x)$ for each x. And the *upper Riemann integral* of $f(x)$ is the largest real number I_{upper} such that

$$I_{\text{upper}} \leqslant \int_a^b \psi(x) \, \mathrm{d}x,$$

where $\psi(x)$ is a step function with $\psi(x) \geqslant f(x)$ for each x. Finally, we say that $f(x)$ is **Riemann integrable** if $I_{\text{lower}} = I_{\text{upper}}$, and call this common value the **Riemann integral** of $f(x)$.

Whilst Riemann's integral was more general than Cauchy's, it still did not assign integrals to unbounded functions and/or functions on unbounded intervals. For example, the integral

$$\int_1^\infty \frac{\mathrm{d}x}{x^2} = 1,$$

which we met earlier, can only be considered by calculating the integral between 1 and X and letting X tend to infinity. The only reasonable answer for the area represented by this integral is 1, but it cannot be evaluated within Riemann's theory and the above is referred to as an *improper Riemann integral*. The limitations of the Riemann integral are discussed further in Chapter 8.

The story of calculus puts paid to any notion that mathematical concepts are conceived complete and polished, or the idea that we are not doing mathematics until the final 'i' is dotted. Indeed, in much of what follows, I will continue referring to infinitesimals as I introduce new types of derivatives and integrals—such language is often the most convenient for giving an informal sense of a concept—but the importance for mathematics of being able to apply calculus rigorously, without reference to ambiguous notions, and to develop general theories of modern analysis, cannot be overstated.

Chapter 4
Should I believe my computer?

Scientists, engineers, and social scientists use mathematics in much of their work, but you may not have considered how the theories of an ideal mathematical world are brought to bear on the real one. A sense of reality can only be achieved via experimentation, but how should we move from a collection of experimental data to wholly defined functions?

Further, the problems met in school classrooms commonly leave a false impression. Such problems usually have exact answers, but with real-world problems, there is typically no hope of finding the *exact* answer.

By way of a first example, consider the following equation:

$$2^{x-2} = x.$$

It's not hard to spot that $x = 4$ is a solution, as $2^{4-2} = 2^2 = 4$. But if we sketch the two graphs $y = x$ and $y = 2^{x-2}$ (Figure 19), then we see that there is a second intersection around where $x = 0.3$. How might we determine the second solution?

We should not be too ambitious—there are deep theorems of mathematics which *prove* that we cannot expect to solve general problems in terms of the so-called 'elementary functions', such as

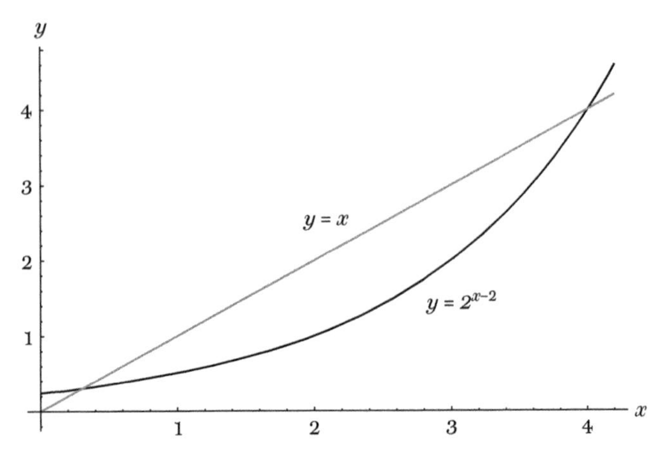

19. Graphs of $y = 2^{x-2}$ and $y = x$.

Table 2. Comparing values of 2^{x-2} and x

x	2^{x-2} (to 4 d.p.)	Comparison
0.2	0.2872	$2^{x-2} > x$
0.3	0.3078	$2^{x-2} > x$
0.4	0.3299	$2^{x-2} < x$
0.31	0.3099	$2^{x-2} < x$

polynomials, exponentials, logarithms, and trigonometric functions—but perhaps an approximate answer, to some desired accuracy, will suffice.

In Table 2 we compare 2^{x-2} and x for some values near $x = 0.3$.

We see that 2^{x-2} is greater than x at $x = 0.2$ and $x = 0.3$, but less than x at $x = 0.4$. The two graphs have crossed somewhere between $x = 0.3$ and $x = 0.4$, meaning the other solution is between 0.3 and 0.4. Trying further values, we see at $x = 0.31$ that 2^{x-2} is less than x, and we now know that the second solution lies

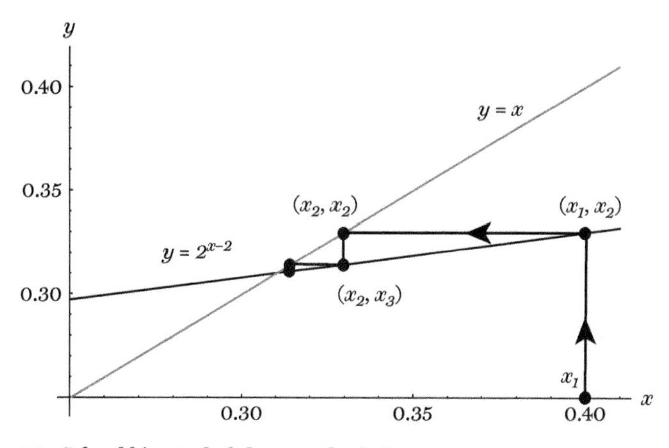

20. Cobwebbing to find the second solution.

between 0.3 and 0.31. We are getting closer, though there is still much work to be done if we want, say, to find the second solution to six decimal places. Can we be more systematic?

Another approach is termed **cobwebbing** (Figure 20). The idea is to start with a nearby estimate of a solution and, ideally, produce a sequence of increasingly accurate approximations. The approach can be applied to solve equations of the form $x = f(x)$, the so-called **fixed points** of the function $f(x)$. In our example, $f(x) = 2^{x-2}$.

We've taken $x_1 = 0.4$ as our initial estimate, as plotted on the x-axis (Figure 20). We then draw a line vertically from $(x_1, 0)$ up to the graph $y = f(x)$ and denote this second point (x_1, x_2) so that $x_2 = f(x_1)$. Moving to the left, we get a third point (x_2, x_2) when we reach the $y = x$ line. (Note that x_2 is closer to the solution than our original estimate x_1.) By repeating this process, generating points (x_2, x_3), (x_3, x_3), (x_3, x_4), ... that lie on the 'descending staircase', we produce a sequence of estimates $x_1, x_2, x_3,$ ... getting ever closer to the solution. In each case, $x_{n+1} = f(x_n)$, and the values of these estimates are given in Table 3.

Table 3. Fixed-point iterations using $f(x) = 2^{x-2}$

x_1	0.4	x_5	0.3101066	x_9	0.3099074
x_2	0.3298770	x_6	0.3099498	x_{10}	0.3099070
x_3	0.3142265	x_7	0.3099161	x_{11}	0.3099070
x_4	0.3108362	x_8	0.3099089	x_{12}	0.3099070

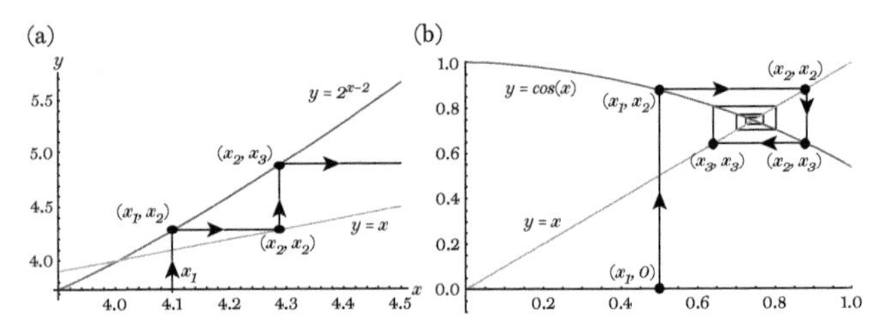

21(a). Cobwebbing near the $x = 4$ solution. 21(b). Cobwebbing when $-1 < f'(x) < 0$.

This fixed-point iteration appears to have provided us with the solution to six decimal places, namely 0.309907, by the ninth estimate x_9; this is certainly fewer steps than our initial approach would have taken. We can verify that this is indeed the solution to six decimal places by comparing 2^{x-2} and x at $x = 0.3099065$ and at $x = 0.3099075$ and showing that $2^{x-2} - x$ changes sign between these values. This is because the x in the range $0.3099065 \leqslant x < 0.3099075$ are precisely those x which round to six decimal places to 0.309907.

The cobwebbing approach converges quickly to the solution (Figure 20), as the gradient $f'(x)$ at the solution is small (the curve is close to horizontal). If we had not immediately spotted that $x = 4$ is a solution, we might have sought to find that solution using cobwebbing. However, we can see (Figure 21(a)) that the estimates $x_1, x_2, x_3, \ldots$ move away from the solution. The problem with this solution is that the gradient $f'(4)$ is greater than 1. A fixed point x is *attracting* if $-1 < f'(x) < 1$. When $-1 < f'(x) < 0$, the estimates $x_1, x_2, x_3, \ldots$ converge to the solution (Figure 21(b)), alternately as over- and underestimates, and the figure looks more like the eponymous cobweb.

This cobwebbing approach highlights some of the important, general characteristics of **numerical analysis**:

- We produce approximations to the exact solution, which can be made as accurate as we wish.
- We have means of checking our answer is correct to the required accuracy.
- Ideally these approximations converge quickly to the solution.

Interpolation and extrapolation

Even the previous problem is idealized, compared with real-world problems: we might not have been able to find the exact solution, but we could at least describe the problem fully. More realistically, the 'functions' associated with a real-world problem won't be so fully specified. In practice, we will just have some experimental data: say a number of experiments are conducted, and the ith experiment outputs y_i when we run the experiment with input x_i. How can we estimate other outputs y when we haven't run the experiment with input x?

Suppose we have a data set from six experiments, as in Table 4.

These six data points (x_i, y_i) are plotted on the graph in Figure 22. There is no y-value for $x = 1.5$, but surely we should be

Table 4. Experimental data points

x_i	1	2	3	4	5	6
y_i	0.3010	0.5490	0.8386	1.2348	1.3632	1.7464

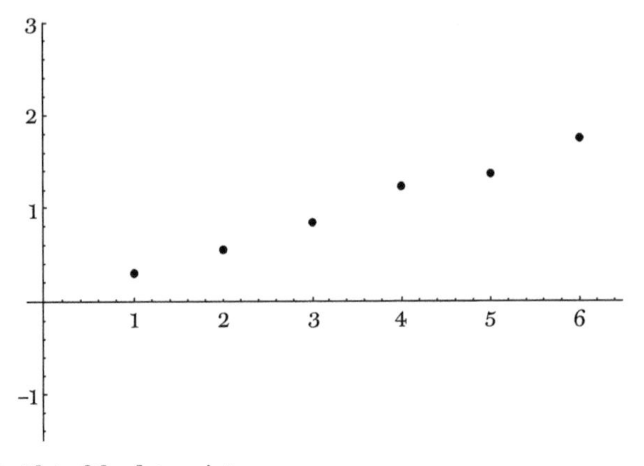

22. Plot of the data points.

able to estimate $y(1.5)$. Our problem, if anything, is that there are many ways to make such an estimate and some academic judgement is needed to decide on the most appropriate method. The process of estimating y-values that correspond to x-values in the given range $1 \leqslant x \leqslant 6$ is known as **interpolation**, and the estimation process outside the given range is called **extrapolation**.

There are various approaches to interpolation, so it would be useful to have some priorities for our estimate $y(x)$ to help choose between the different methods. It would be convenient if:

- the formula for $y(x)$ is relatively simple and uncomplicated to evaluate;
- the function agrees (or almost agrees) with the data points (x_i, y_i);
- the function does not fluctuate wildly between the data points;
- the function $y(x)$ is differentiable at least once, perhaps more often;
- the function has a form which is plausible, given the nature of the experiment.

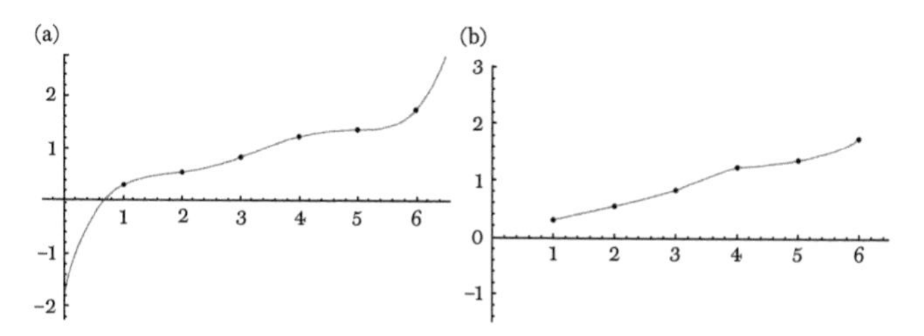

23(a). Polynomial interpolation. 23(b). Interpolation with splines.

Polynomial (or Lagrangian) interpolation: Given n data points, there is a unique polynomial of degree less than n whose graph passes through the n points. For the given example, this is the polynomial of degree five plotted (Figure 23(a)). Two weaknesses with this approach are that a degree five polynomial is relatively complicated to calculate with, and the polynomial extrapolates poorly beyond the range $1 \leqslant x \leqslant 6$.

Splines: One way around the problem of using a single high-degree polynomial is to use a number of *different* polynomials of low degree to interpolate the data (Figure 23(b)). For example, a *cubic spline* is a function $y(x)$ such that:

- $y(x)$ is defined by some cubic (degree 3) polynomial $p_i(x)$ between x_i and x_{i+1};
- $y(x_i) = y_i$ at each data point;
- $y(x)$ has a continuous second derivative.

Least-squares approximation: The data points plotted (Figure 22) appear to lie, approximately, on a straight line. No single line passes through all six points, but this is to be expected, as the data were generated experimentally and there will always be some experimental error. But is there a 'best-fit' line in some sense? The equation of a line is $y = ax + b$, so for any such line the model's estimate of $ax_i + b$ differs a little from the experimentally recorded

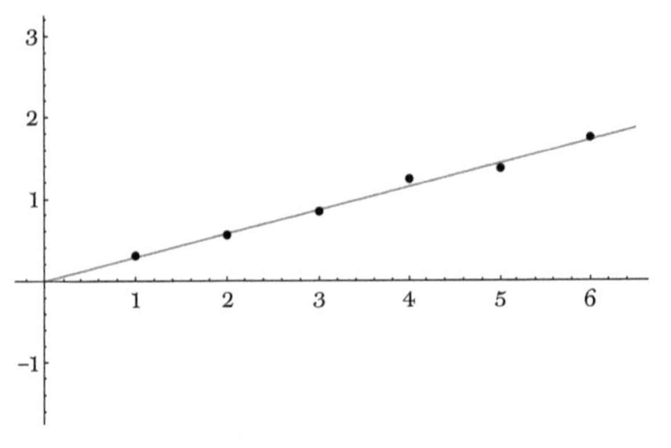

24. Linear approximation.

answer of y_i. The **least-squares approximation** is the line $y = ax + b$, for which the total error

$$E(a, b) = (ax_1 + b - y_1)^2 + (ax_2 + b - y_2)^2 + \cdots + (ax_6 + b - y_6)^2$$

is minimized. The error $E(a, b)$ is a function of a and b and can be minimized using multivariable calculus (Chapter 5). For the given data the best-fit line (Figure 24) is when

$$a = 0.2876 \text{ and } b = -0.0011.$$

The actual data in Table 4 was produced using $y_i = (ax_i + b) + \varepsilon$, where

$$a = 0.3126 \text{ and } b = -0.1416$$

and where ε represents a random error from the range $-0.15 \leqslant \varepsilon \leqslant 0.15$. These estimates for a and b are not particularly accurate here, but there are clear reasons for this: practically, we would hope for more than six data points to work with; moreover, an experimental error of up to 0.15 is substantial for outputs y_i in the range 0 to 2. For better estimates we would need to collect more data and/or improve our experiment to reduce the errors.

Least-squares approximation can be used for families of curves other than straight lines. It may be that the model for an experiment implies that a solution should have the form

$$a \sin(x) + b \cos(x)$$

for some a, b yet to be determined. A different error formula $E(a, b)$ could be created as before and minimized, but now using terms such as $\left(a \sin(x_i) + b \cos(x_i) - y_i\right)^2$.

Numerical differentiation and integration

In 1768 Euler presented a method for finding approximate solutions to the initial value problem

$$\frac{dy}{dx} = f(x, y), \qquad y(x_0) = y_0.$$

If we know a point (x, y) on the solution's graph, then the above differential equation also tells us the gradient of the solution at that point, namely $f(x, y)$. The premise of Euler's method is that for a suitably small increment h in x, the solution's graph is approximated well by a line segment with one endpoint at (x, y) and gradient $f(x, y)$ (Figure 25). If the increment in x is h, then the other endpoint is at $(x + h, y + hf(x, y))$.

We know for certain that the solution's graph has a point (x_0, y_0) on it. Working from that point, we can then generate the next estimate (x_1, y_1) where

$$x_1 = x_0 + h, \qquad y_1 = y_0 + f(x_0, y_0)h,$$

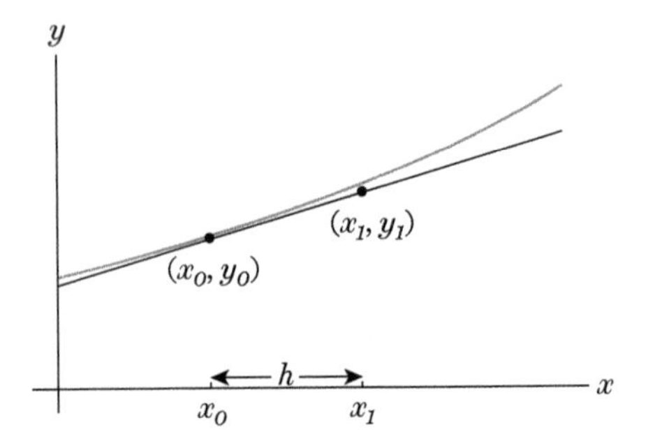

25. Euler's method—the first iteration.

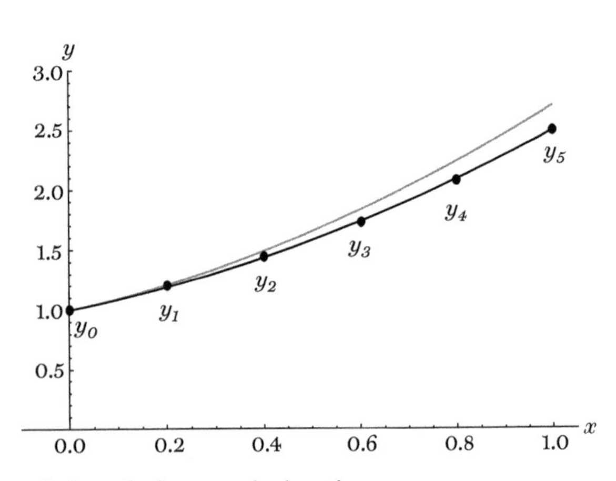

26. Euler's method—successive iterations.

and then (Figure 26) generate subsequent estimates y_n (for $n \geqslant 1$) by using

$$x_n = x_0 + nh, \qquad y_n = y_{n-1} + f(x_{n-1}, y_{n-1})h.$$

Recall that the exponential function $y = e^x$ is the solution to the initial value problem $y' = y$ and $y(0) = 1$, so that $f(x, y) = y$.

If we were to apply Euler's method to this problem, then we would have $x_n = nh$, generating the estimates

$$y_0 = 1; \qquad y_1 = 1 + h; \qquad y_2 = 1 + h + (1 + h)h = (1 + h)^2,$$

and, generally, we would find that

$$y_n = (1 + h)^n = \left(1 + \frac{x_n}{n}\right)^n.$$

If we fix $x = nh$ but use decreasing increments h, or equivalently increasing n, then the estimate of $y(x)$ converges to the solution e^x, as this is Harriot's original definition of the exponential.

Note that the formula from Euler's method can be rearranged as

$$\frac{y_n - y_{n-1}}{h} = f(x_{n-1}, y_{n-1}),$$

which is a natural approximation of the differential equation $y' = f(x, y)$. In fact, such an approximation for y' can be used more generally for other differential equations. Similar approximations exist for higher-order derivatives—for example,

$$\frac{y_n - 2y_{n-1} + y_{n-2}}{2h^2}$$

provides an approximation for the second derivative y''.

Since Euler's time, his method has been variously improved upon, and would not be widely used today, but it gives a sense of how one might seek approximate solutions to a differential equation when no exact solution is possible. One way in which the method is simplistic is that the same increment h is used throughout. However, the function $f(x, y)$ may change rapidly for some values of x, compared with others. It would make sense to use much smaller increments when this is the case.

For other problems, it may be imperative that y is a periodic function—for example, if it describes the distance to a planet which traces the same orbit each year. Whatever approximating method is used, the approximate solution must be periodic for it to make sense physically.

There are also various means for estimating definite integrals. The function $y = \sin(x^2)$ for x in the interval $0 \leqslant x \leqslant \pi/2$ is graphed (Figures 27, 28). The value of the desired integral is

$$\int_0^{\pi/2} \sin(x^2)\mathrm{d}x = 0.828116\ldots$$

First the **trapezium rule** is employed (Figure 27), dividing the original interval into four. In general, the trapezium rule splits the given interval into equally wide subintervals. An approximation to the graph of $f(x)$ is created by connecting the points $(x_i, f(x_i))$ with lines. The approximating graph overlies trapezia, whose total (signed) area gives an approximation of the integral.

Simpson's rule (Figure 28) is a little more refined. An even number of subintervals needs to be used—here four subintervals

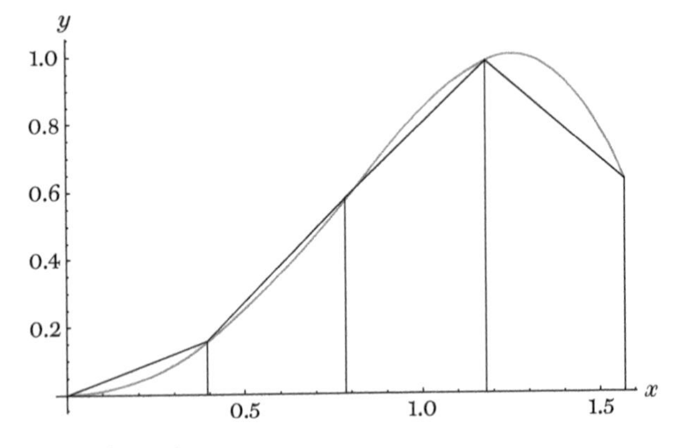

27. Trapezium rule.

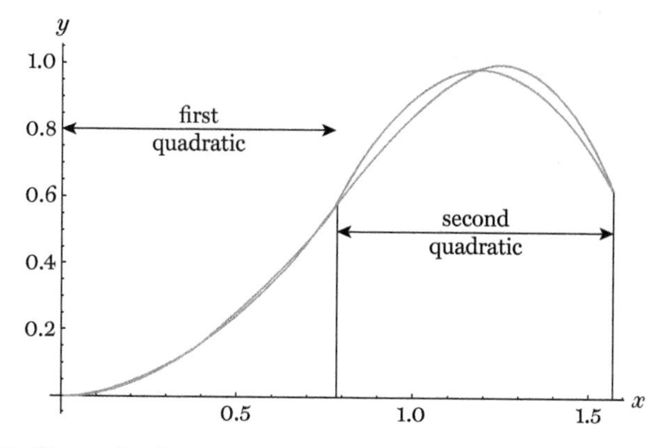

28. Simpson's rule.

are used, but they are grouped as x_0, x_1, x_2, and x_2, x_3, x_4. Simpson's rule approximates $f(x)$ between x_0 and x_2 using the unique quadratic that passes through the first three points $(x_0, f(x_0)), (x_1, f(x_1)), (x_2, f(x_2))$; a different quadratic is then used for the interval between x_2 and x_4, and so on. Simpson's rule gives an estimate for the integral of $f(x)$ by integrating these quadratics instead.

These two rules, using four subintervals, give estimates of 0.79621 and 0.82845 to five decimal places. In general, Simpson's rule produces better estimates. The rule is named after Thomas Simpson, and appears in his calculus text of 1743. But this was a 'rediscovery'; it had been known to mathematicians as early as the 17th century.

In all this, note that there are pros and cons to using smaller increments. Yes, the accuracy of our estimate will improve, but the computational time and memory needed will also increase. Depending on the specifics of a problem or experiment, an estimate will only be required to a certain accuracy, and it would be largely pointless to put resources into obtaining further accuracy.

Numerical stability and error analysis

Numerical stability relates to the concern that some algebraic operations may exacerbate errors or approximations of values. A value used in a computation may differ from a notional 'true' value either because the value was measured experimentally or because the value was the true value rounded to a certain number of decimal places. Such stability concerns are particularly important in a long computation involving thousands of iterations: the issue of rounding errors may lead to nonsensical answers being generated.

If x denotes a true value and x_1 approximates x, then the difference in outputs $f(x)$ and $f(x_1)$ is approximately

$$f'(x_1)(x - x_1),$$

provided $f(x)$ is differentiable. So, approximate errors won't be exacerbated if the derivative $f'(x)$ is small. This concurs with the behaviour we saw earlier with cobwebbing—recall that a fixed point a of $f(x) = x$ is attracting if the derivative $f'(a)$ lies between -1 and 1. The error improves by a factor of approximately $f'(a)$ with each iteration.

For some iterations we can expect even faster convergence. **Newton's method** is a means of approximating solutions of an equation $f(x) = 0$. The idea for the iteration is captured in Figure 29. If we have a first approximation x_1 to the solution a, we might expect that

$$x_2 = x_1 - \frac{f(x_1)}{f'(x_1)}$$

is a better approximation. This value of x_2 occurs where the tangent line to $y = f(x)$ at the point $\left(x_1, f(x_1)\right)$ crosses the x-axis. (See the Appendix for details.)

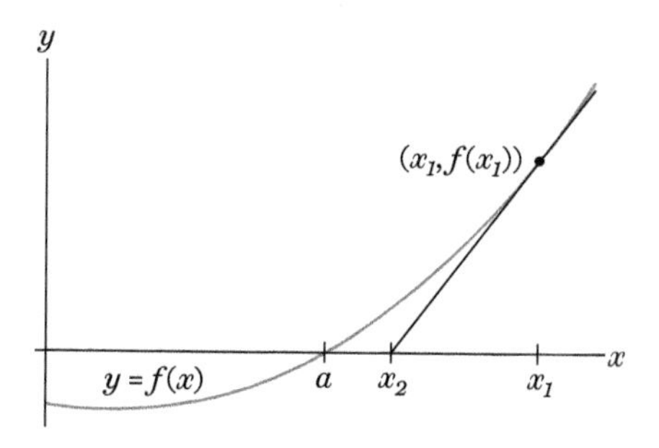

29. Newton's method.

We can then replace x_1 with x_2 to produce the next iteration x_3, and so on. Under technical but broad conditions—essentially for a graph that looks like the one plotted (Figure 29)—the iterations x_n will converge as a decreasing sequence to the solution a. Even better, this convergence is *quadratic*; this means that there is a positive constant M such that

$$|x_{n+1} - a| \leqslant M|x_n - a|^2.$$

The expression $|x_n - a|$ is the error of the nth iteration x_n from the solution a. This will be a small number as the iteration progresses, so its square $|x_n - a|^2$ will be yet smaller. Such quadratic convergence will be much faster than the linear convergence achieved by cobwebbing.

Maintaining some bounds on errors is an important part of numerical analysis, and it provides certainty that the approximations work to the required accuracy. For Euler's method it can be shown that the error is proportional to the increment h, and so the estimates will converge to the correct solution as h decreases. Similarly, it can be shown that the error for the trapezium rule with n intervals is proportional to $1/n^2$, while the error for Simpson's rule is proportional to $1/n^4$. This shows that Simpson's rule is superior.

Linearization and stability

The motion of a swinging pendulum can be modelled by the differential equation

$$\frac{d^2\theta}{dt^2} = -\frac{g}{l}\sin(\theta).$$

A pendulum consists of a light rod of length l with a mass at its end; here g denotes acceleration due to gravity and $\theta(t)$ denotes the angle the pendulum makes with the downward vertical at time t (Figure 30(a)).

If the pendulum starts from rest at an angle α, then this initial value problem can be largely solved to show that the pendulum swings with period

$$T = 4\sqrt{\frac{l}{2g}}\int_0^\alpha \frac{1}{\sqrt{\cos(\theta) - \cos(\alpha)}}\,d\theta.$$

Unfortunately, this integral cannot be evaluated exactly. However, if the pendulum makes only small oscillations, then

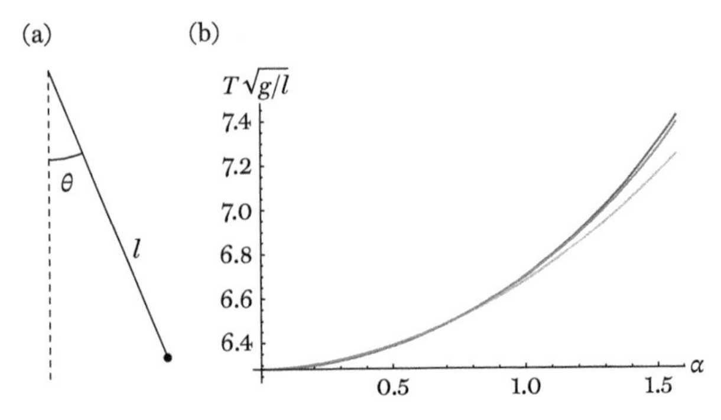

30(a). Pendulum. 30(b). Period of a pendulum.

we can use the trigonometric series to approximate the pendulum's motion.

Recall the sine and cosine series:

$$\sin(\theta) = \theta - \frac{\theta^3}{3!} + \frac{\theta^5}{5!} - \frac{\theta^7}{7!} + \cdots, \qquad \cos(\theta) = 1 - \frac{\theta^2}{2!} + \frac{\theta^4}{4!} - \frac{\theta^6}{6!} + \cdots.$$

If θ is small, then $\theta^2, \theta^3, \theta^4, \theta^5, \ldots$ become ever smaller. So, for suitably small angles, $\theta - \theta^3/6$, or even just θ, will be a reasonable approximation for $\sin(\theta)$. If we use the approximation $\sin(\theta) \approx \theta$ in the original differential equation, we get $\theta''(t) = (-g/l)\theta$, and this *is* an initial value problem that we can solve, obtaining

$$\theta(t) = \alpha \cos\left(\sqrt{\frac{g}{l}}t\right).$$

Note that the period of this function, $2\pi\sqrt{l/g}$, does not depend on the initial angle α, provided that α is small. This is a good first approximation to T, accurate to one per cent for α up to $20°$, but we might use more terms from the sine or cosine series to produce yet better approximations, such as the power series

$$T = 2\pi\sqrt{\frac{l}{g}}\left(1 + \frac{\alpha^2}{16} + \frac{11\alpha^4}{3072} + \cdots\right).$$

Now (Figure 30(b)) $T\sqrt{g/l}$ as a function of α is the top graph plotted alongside the approximations for $T\sqrt{g/l}$ using the series approximation up to the α^2 term (bottom graph) and also up to the α^4 term (middle graph); this last approximation is almost indistinguishable from the true graph.

Earlier, when we replaced $\sin(\theta)$ by θ, we were performing **linearization**. This involves treating all powers of degree higher than one as negligible. Linearization can help study the **equilibria** of physical systems. An equilibrium is a state of a system where, in principle at least, the system might remain at rest. For a pendulum, such an equilibrium has it hanging downwards. The previous discussion showed this equilibrium to be **stable** and determined the period of small oscillations about the equilibrium. Another equilibrium for the pendulum is when it hangs upright—this is theoretically possible, albeit precarious. We can linearize the differential equation about the vertical and see that realistically the pendulum will move exponentially away from the upward vertical—that is, this equilibrium is unstable.

Linearization is an important tool of analysis, especially for the study of equilibria, but it fails to capture the qualitative global nature of general systems. A system might exhibit **chaos**, meaning that the system can evolve in vastly different ways from two very close starting positions—the so-called *butterfly effect*. Chaos is seen in a system as simple as the *double pendulum*—where a second pendulum is attached to the bottom of a pendulum. Consequently, chaos causes great challenges for numerical methods.

The next chapter introduces the calculus of functions of more than one variable, to which the numerical and approximating techniques that have been described here can be extended. For example, in the previous chapter (Figure 15(b)), we saw that there are two equilibria for the Lotka–Volterra model: one is when $F = R = 0$, so that both populations are absent/extinct and this is an unstable equilibrium; the second equilibrium is when $F = b/k$ and $R = m/a$; this is the point at the centre of the egg-shaped cycles. This equilibrium is stable: if the system is at a nearby point, linearization can be used to show that the point (F, R) cycles on a small egg-shaped curve around the equilibrium with a period of $2\pi/\sqrt{bm}$ (see the Appendix for details).

Chapter 5
Dimensions aplenty

In Chapter 2 we introduced the calculus of real functions of one variable—that is, functions that take one numerical input and produce one numerical output. Most functions, though, are not of such a simple form. As you read this, in this instant, the temperature T around you might be described as a function of three spatial co-ordinates x, y, z needed to describe a point's position; $T(x, y, z)$ would then denote the current temperature at that point. This function T has three numerical inputs and one numerical output. It seems physically reasonable that T should be both continuous and differentiable, but perhaps it is not quite clear what it means for a function $T(x, y, z)$ with three inputs to be differentiable—we certainly can't just graph this function in the plane and define the derivative as the gradient of a tangent line.

Despite this, you may have heard the phrase 'temperature gradient' used. Informally, this might relate to the change in temperature as you move in a particular direction. If the temperature gradient were negative, then you would be getting colder, and if positive, getting hotter. More precisely, the temperature gradient is a *vector*—its magnitude is the greatest rate of increase in temperature over all directions, and its direction is how you would need to move to appreciate that greatest rate of increase.

Scalars and vectors

In mathematics and science, there is an important distinction between **scalar** quantities and **vector** quantities. Temperature is a scalar quantity, meaning that the output $T(x, y, z)$ is a single real number. Many quantities are scalar, including distance, speed, angle, density, mass, volume, energy, charge, power, and probability. All of these can be represented by a single real number, which may be positive, zero, or negative in particular cases.

However, many quantities cannot be represented in this way—examples include wind velocity, gravitational force, electromagnetic fields, and angular velocity. It may be the case that the wind is blowing at five metres per second—this is the speed of the wind or its velocity's *magnitude*—but to fully describe the wind's velocity, we would need to describe the wind's *direction* as well: to what extent it is blowing forwards or backwards, left or right, up or down. Quantities that have a magnitude and a direction are called vectors.

Once we have introduced three spatial co-ordinates x, y, z, a vector quantity can be represented by three *components* (v_x, v_y, v_z). Such a vector might also be denoted by a single letter in bold such as **v**. If **v** were to represent wind velocity, then v_z would be the extent to which the wind was blowing up (if v_z is positive) or down (if v_z is negative). And the magnitude of **v** is denoted as $|\mathbf{v}|$, which is given by the formula

$$|\mathbf{v}| = \sqrt{(v_x)^2 + (v_y)^2 + (v_z)^2}.$$

If **v** represents wind velocity, then $|\mathbf{v}|$ is the wind speed. If a vector $\mathbf{r} = (x, y, z)$ represents the position of a point from an origin $(0, 0, 0)$ (Figure 31(a)), then, by Pythagoras' theorem, $|\mathbf{r}|$ is the length of the vector **r**, or equally the distance of the point (x, y, z)

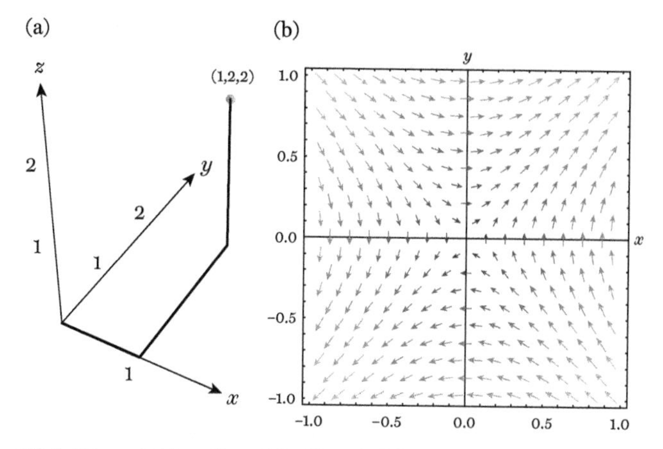

31(a). The point (1,2,2). 31(b). The wind flow v(x, y) = (y, x).

from the origin. A two-dimensional wind flow is shown in Figure 31(b).

Finally, we need to be clear what we mean by direction. If a wind with velocity **v** blows in a certain direction, then a second wind with velocity **2v** blows in the *same* direction but with *twice* the speed. Consequently, we will use **unit vectors** to describe directions—that is, vectors of magnitude (or length) one.

Directional and partial derivatives

Let $T(x, y, z)$ be the temperature T at a point (x, y, z). Depending how we move from this point, we might get colder or warmer. Using vectors, we now have an appropriate language to describe such movements. If we move in a certain direction, represented by a unit vector **u**, then the **directional derivative** of T in the direction **u** is the rate of change of temperature T as we move in direction **u**.

There are three particularly important directional derivatives, the so-called **partial derivatives**, which correspond to the direction **u** being parallel to one of the three co-ordinate axes. When $\mathbf{u} = (1,0,0)$, so that the direction is parallel to the x-axis, the directional derivative is denoted by

$$\frac{\partial T}{\partial x},$$

and called the 'partial derivative of T with respect to x'. It is read as 'partial d T by d x'. This is the rate of change in T as we increase x while keeping y and z constant. There are similar partial derivatives $\frac{\partial T}{\partial y}$ when $\mathbf{u} = (0,1,0)$ and $\frac{\partial T}{\partial z}$ when $\mathbf{u} = (0,0,1)$.

For example, if $T(x,y,z) = 3x^2 + 2y^2 + z^2 + 2xz$, then

$$\frac{\partial T}{\partial x} = 6x + 2z, \qquad \frac{\partial T}{\partial y} = 4y, \qquad \frac{\partial T}{\partial z} = 2z + 2x,$$

as in each case we differentiate T with respect to the given variable (x or y or z) and treat the other two variables as if they are constants.

A general directional derivative can be expressed in terms of the partial derivatives. If $\mathbf{u} = (u_1, u_2, u_3)$ is a unit vector, then the directional derivative of T in the direction $\mathbf{u}$ equals

$$u_1\frac{\partial T}{\partial x} + u_2\frac{\partial T}{\partial y} + u_3\frac{\partial T}{\partial z}.$$

It follows that the vector

$$\mathrm{grad}(T) = \left(\frac{\partial T}{\partial x}, \frac{\partial T}{\partial y}, \frac{\partial T}{\partial z}\right)$$

is in the direction where T increases fastest. This vector $\mathrm{grad}(T)$ is called the **gradient vector** of T, usually read 'grad T'.

For the earlier function T, we have $\text{grad}(T) = (6x + 2z, 4y, 2z + 2x)$. So at the point $(1, 0, -1)$, the temperature is increasing fastest parallel to $(4, 0, 0)$, which has direction $(1, 0, 0)$.

Recall Fermat's theorem: if a function $f(x)$ is maximal or minimal, then $f'(x) = 0$. In a similar manner, if $T(x, y, z)$ is maximal or minimal, then

$$\frac{\partial T}{\partial x} = \frac{\partial T}{\partial y} = \frac{\partial T}{\partial z} = 0 \text{ or, equivalently, } \text{grad}(T) = (0, 0, 0) = \mathbf{0}.$$

For the given example, $\text{grad}(T) = \mathbf{0}$ at $(x, y, z) = (0, 0, 0)$, which is actually a minimum for the function T. In the Appendix we show how the least-squares error $E(a, b)$ from Chapter 4 can be minimized to find the best-fit line.

As with 'full' differentiation, second (and higher) partial derivatives can be formed by repeated partial differentiation. Under mild technical requirements, the order of differentiation won't matter; for example, using the earlier function $T(x, y, z)$, we obtain the same answer whether we differentiate with respect to z first then x or vice versa:

$$\frac{\partial^2 T}{\partial x \partial z} = \frac{\partial}{\partial x}\left(\frac{\partial T}{\partial z}\right) = \frac{\partial}{\partial x}(2z + 2x) = 2,$$
$$\frac{\partial^2 T}{\partial z \partial x} = \frac{\partial}{\partial z}\left(\frac{\partial T}{\partial x}\right) = \frac{\partial}{\partial z}(6x + 2z) = 2.$$

Partial differentiation equations

We saw in Chapter 3 that the development of calculus was intimately linked with differential equations. In that chapter we were interested in *ordinary* differential equations (ODEs)—that is, ones involving a function $f(x)$ of a single input x. As examples, the ODEs

$$\frac{\mathrm{d}f}{\mathrm{d}x} = 0 \qquad \text{and} \qquad \frac{\mathrm{d}^2 f}{\mathrm{d}x^2} = 0$$

have general solutions

$$f(x) = c \qquad \text{and} \qquad f(x) = c_1 x + c_2,$$

where $c,\ c_1,$ and c_2 are constants (that is, real numbers). Provided a version of Picard's theorem applies, the general solution to an order n ordinary differential equation involves n arbitrary constants.

In a certain sense, the above also applies to **partial differential equations** (PDEs)—differential equations involving partial derivatives. The first-order PDE

$$\frac{\partial f}{\partial x} = 0,$$

involving a function $f(x, y)$ of two variables x, y, has the general solution $f(x, y) = c(y)$. Here $c(y)$, rather than being an arbitrary constant, is an arbitrary function of y. As the partial derivative $\partial f / \partial x$ equals 0, $f(x, y)$ remains constant as x varies or, put another way, $f(x, y)$ can vary only with the other variables; here the only other variable is y.

Likewise, the second-order PDEs

$$\frac{\partial^2 f}{\partial x^2} = 0, \qquad \frac{\partial^2 f}{\partial x \partial y} = 0, \qquad \frac{\partial^2 f}{\partial x^2} = \frac{\partial^2 f}{\partial y^2}$$

respectively have solutions

$$f(x, y) = c_1(y)x + c_2(y),$$
$$f(x, y) = c_1(x) + c_2(y),$$
$$f(x, y) = c_1(x - y) + c_2(x + y),$$

where c_1 and c_2 are arbitrary functions.

The study of PDEs is a substantial field of mathematics, with such equations found ubiquitously across science. We shall meet some important PDEs in the next two chapters.

The calculus of variations

A good deal of calculus is about optimization—what is the least or greatest that something can be. When that something depends on only one input, Fermat's theorem tells us that the derivative has to equal zero. When there are several inputs, the partial derivatives need to be zero, but some optimization problems require more complicated mathematics than this.

For example: what is the curve of shortest distance between two points in the plane? The correct answer is the line segment connecting the two points. But it takes considerable thought to understand what is being claimed here and what needs proving.

By Pythagoras' theorem, the two points $(0,0)$ and $(1,1)$ in the plane are distance $\sqrt{2}$ apart, this being the length of the line segment between them. Our claim is that this distance is less than the length of *any* other curve connecting the two points.

Several such curves are sketched in Figure 32. For a general curve $y = f(x)$, between $(0,0)$ and $(1,1)$, its length equals

$$\int_0^1 \sqrt{1 + \left(f'(x)\right)^2}\, \mathrm{d}x.$$

To each such curve we have assigned an integral and we seek to find the smallest of all these integrals; here we claim this to be when $f(x) = x$, as $y = x$ is the equation of the line connecting the points. This is still clearly an optimization problem, but

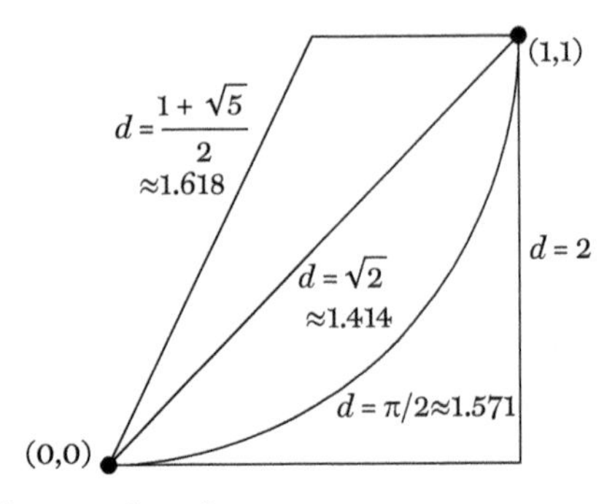

32. Curves between the points.

mathematically one of much more sophistication than any previously met.

Such an integral is minimal when $F(x, f, f')$, the function being integrated, satisfies the **Euler–Lagrange equation**,

$$\frac{\mathrm{d}}{\mathrm{d}x}\left(\frac{\partial F}{\partial f'}\right) = \frac{\partial F}{\partial f}.$$

In the given example we have $F(x, f, f') = \sqrt{1 + (f')^2}$, and it's relatively straightforward to show that $f(x) = x$ satisfies the Euler–Lagrange equation in this case (see Appendix). The more general case for curves between any two points can be similarly demonstrated.

Further, the Euler–Lagrange equation can be used to find the shortest curves on surfaces, known as *geodesics*—for example, the great circles on a sphere are geodesics (a great circle cuts a sphere into two hemispheres)—and light travels along geodesics in the theory of general relativity.

These problems are typical of the **calculus of variations**. A famous early question was the **brachistochrone** problem: given two

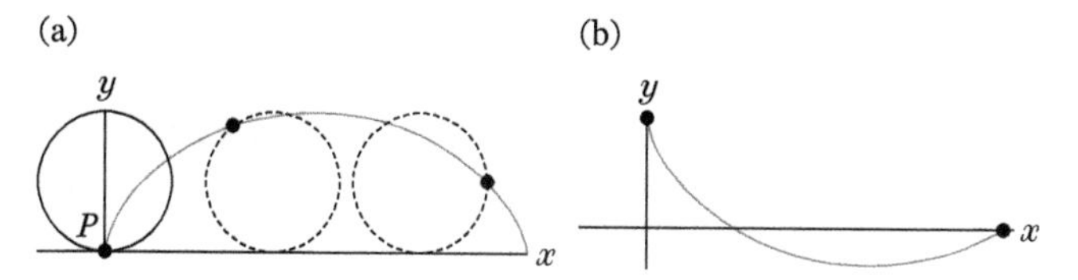

33(a). A cycloid. 33(b). A solution to the brachistochrone.

points, one higher than the other (Figure 33(b)), what is the shape of a smooth wire, connecting the two points, so that a bead takes the shortest time possible, sliding from the higher point to the lower under gravity? The problem's name 'brachistochrone' derives from the Greek for 'shortest time'.

This problem was posed in 1696 by Johann Bernoulli, who himself knew the answer. He received correct solutions from his brother Jacob, Leibniz, and Newton. The answer is not, as you might guess, the straight line between the two, but rather an upside-down **cycloid.**

The cycloid is a curve of geometric interest, as it is the curve traced out by a fixed point P on the circumference of a rolling circle (Figure 33(a)). The solution to the brachistochrone problem is the upside-down cycloid beginning vertically and passing through the two points, and this problem can be solved using the Euler–Lagrange equation. In this case the integral associated with each function is the time taken for a particle to move along a smooth wire in the shape of the function's graph.

A similar problem is the *isoperimetric problem*: given a loop of string, what's the largest region you can encompass with it? You might guess the answer is a circle, and you'd be right—any optimizing curve to be a circle. But this wasn't proven until 1870 by Weierstrass. Specifically, he proved the **isoperimetric inequality**: if the string has length L, and bounds a region of area A, then

$$L^2 \geqslant 4\pi A,$$

with $L^2 = 4\pi A$ holding only for circles (when $L = 2\pi r$ and $A = \pi r^2$, where r is the circle's radius). So given string of length L, the greatest area it can bound is $L^2/(4\pi)$.

The same question can be posed in higher dimensions, and in one dimension higher the optimizing surfaces are spheres. This is the reason that *soap bubbles* are spherical. Because of energy considerations, the soap bubble seeks to reduce its surface tension, and so area, while containing the same volume of air. Likewise, a *soap film* spanning a wire boundary seeks to minimize its area, and hence the surfaces that soap films make are examples of **minimal surfaces**. For example, given two parallel circular boundaries, the soap film forms a *catenoid* (Figure 34).

In 1760, Lagrange first posed the following problem: given a bounding wire, is there a minimal surface having that boundary?

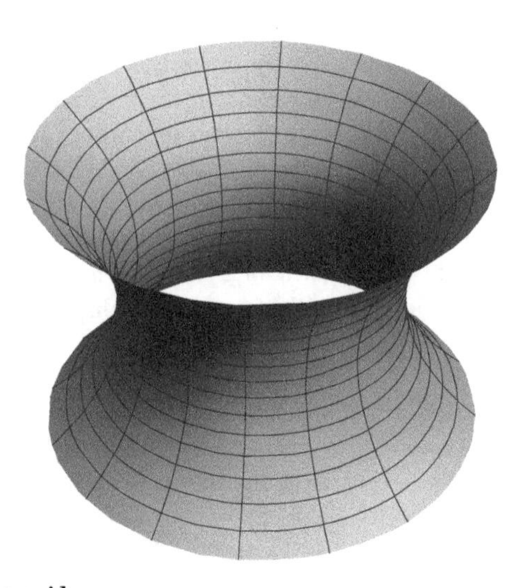

34. A catenoid.

Physically, this seems intuitively clear, as it's whatever form a soap film would take spanning the wire; the problem is now known as **Plateau's problem** after the physicist Joseph Plateau, who investigated this problem using soap films. However, a mathematical proof of the existence of smooth minimal surfaces for a general boundary involved surmounting significant technical difficulties and was not completed until 1970.

Multivariable integration

Just as there is a differential calculus of several variables, so there is an integral calculus too. In one variable, integration is a means of rigorously defining a sum of infinitesimal contributions and calculating (signed) areas under a graph. But integrals can represent other quantities: as examples, if we integrate a velocity, a displacement is evaluated, and integrals can also represent probabilities.

If a function $f(x)$ is positive, then the integral $\int_a^b f(x)\,\mathrm{d}x$ represents the area below the graph $y = f(x)$ and above the interval $a \leqslant x \leqslant b$. Given a positive function $f(x, y)$ of two variables, its graph $z = f(x, y)$ is a surface above the xy-plane, and it seems reasonable that some integral should represent the volume under this surface and above a region R of the plane. This volume is denoted by

$$\iint_R f(x, y)\,\mathrm{d}A$$

(Figure 35(a)). Previously, we informally thought of $f(x)\,\mathrm{d}x$ as an infinitesimally thin rectangle of height $f(x)$ and base $\mathrm{d}x$. Now $f(x, y)\,\mathrm{d}A$ is the volume of a prism, of height $f(x, y)$ and infinitesimally small base of area $\mathrm{d}A$ in the xy-plane.

If the set R is a complicated region, then subtle techniques may be needed to calculate this volume, but when R is a rectangle—a

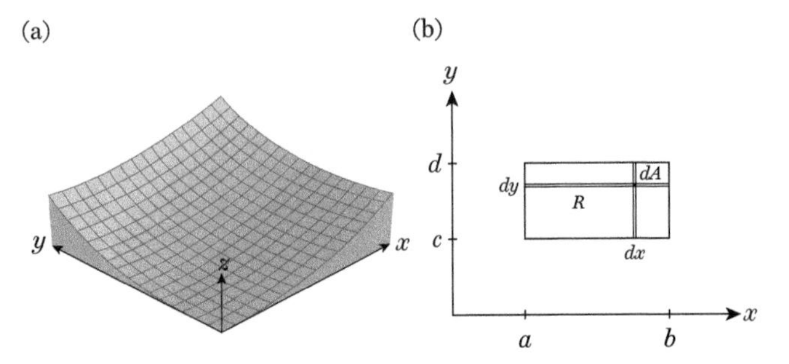

35(a). Volume under a surface above a square. 35(b). Rectangle.

natural generalization of an interval—such volumes can be readily evaluated. If R is the rectangle

$$a \leqslant x \leqslant b, \quad \text{and} \quad c \leqslant y \leqslant d$$

(Figure 35(b)), then the integral can be determined as the 'repeated' or 'double' integral

$$\int_{x=a}^{x=b} \left(\int_{y=c}^{y=d} f(x, y) \, \mathrm{d}y \right) \mathrm{d}x.$$

As an example, the volume shown above (Figure 35(a)), which lies under the surface $z = x^2 + y^3$ and above the square $0 \leqslant x, y, \leqslant 1$, equals

$$\int_{x=0}^{x=1} \left(\int_{y=0}^{y=1} (x^2 + y^3) \, \mathrm{d}y \right) \mathrm{d}x.$$

To evaluate the middle y-integral, we treat x as a constant and note that $x^2 y + y^4/4$ is an antiderivative of $x^2 + y^3$. By the fundamental theorem,

$$\int_{y=0}^{y=1} (x^2 + y^3) \, \mathrm{d}y = \left(x^2 \times 1 + \frac{1^4}{4} \right) - \left(x^2 \times 0 + \frac{0^4}{4} \right) = x^2 + \frac{1}{4}.$$

This has an antiderivative $x^3/3 + x/4$, and so, by the fundamental theorem, the volume is

$$\int_{x=0}^{x=1} \left(x^2 + \frac{1}{4}\right) \mathrm{d}x = \left(\frac{1^3}{3} + \frac{1}{4}\right) - \left(\frac{0^3}{3} + \frac{0}{4}\right) = \frac{1}{3} + \frac{1}{4} = \frac{7}{12}.$$

Volumes aren't the only things represented by such integrals. If the above square were made of a material with density $f(x,y) = x^2 + y^3$, then $\frac{7}{12}$ would be the mass of that square. When $f(x,y) = 1$, then $\iint_R f(x,y)\,\mathrm{d}A$ equals the area of R. If the region R were a disc representing a dartboard, and $f(x,y)\,\mathrm{d}A$ denoted the infinitesimal probability of a dart landing at a point (x,y), then $\iint_R f(x,y)\,\mathrm{d}A$ would be the probability of the dart player hitting the dart board—we would expect this to equal 1 (representing certainty), or just below 1, given the occasional miss. If $s(x,y)$ were the score received for a dart landing at (x,y), then $\iint_R s(x,y)f(x,y)\,\mathrm{d}A$ would be the average score achieved with a dart. These ideas extend to three and higher dimensions.

Note there is a natural order when we sum an integral $\int_a^b f(x)\,\mathrm{d}x$; we think of x as increasing, varying from $x = a$ up to $x = b$. With a rectangular region R, there are two natural ways to cover the rectangle: either vary y first and then x (as we did above) or vice versa. For more complicated regions, there may not seem any preferential way to let (x,y) vary over R. In some instances, the order of integration can matter, as Riemann showed for infinite sums. But if the modulus $|f(x,y)|$ is integrable on R, then the integral $\iint_R f(x,y)\,\mathrm{d}A$ always gives the same value, irrespective of the order in which it is summed. In 1837 Dirichlet demonstrated the same for infinite sums: if the sum of the modulus of the terms converges, then the infinite sum always gives the same answer, no matter how it's rearranged.

Surface integrals and flux

A multivariable integral can represent the area of a region in the plane, but we might wish to calculate the area—or **surface**

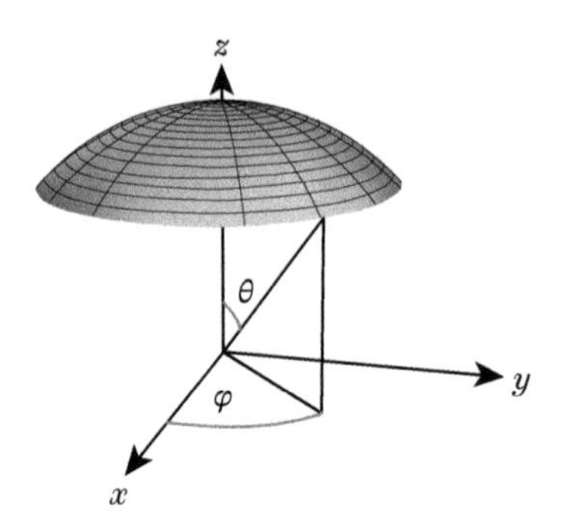

36. Latitude and longitude.

area—of a non-planar surface, such as a sphere or a cone situated in three-dimensional space. You may even know such formulae already—for example, the area of a sphere of radius a is $4\pi a^2$.

Our first problem is how we might describe a surface like a sphere, and one way to do this is to assign co-ordinates to the surface. For a sphere we might use *latitude* and *longitude*. We can measure the angle θ of latitude from the north pole, where $\theta = 0$; past the equator, where $\theta = \pi/2$; or to the south pole, where $\theta = \pi$ (Figure 36). Similarly, we can measure the angle φ of longitude from the prime/Greenwich meridian, going once around the world so that $\varphi = 2\pi$ when we return to Greenwich. (Note we are again using radians here, and the ranges of θ and φ are different to those traditionally used on a map or globe.)

The area of a planar region R is $\iint_R \mathrm{d}A$. Informally, we can think of this as a sum of infinitesimal rectangles of area $\mathrm{d}A = \mathrm{d}x\ \mathrm{d}y$, where the point (x, y) ranges over R (Figure 35(b)). To determine the surface area of the sphere, we need to sum infinitesimal elements $\mathrm{d}S$ of surface area, as the co-ordinates range over all possible values. However, it is not simply the case that $\mathrm{d}S = \mathrm{d}\theta\ \mathrm{d}\varphi$ on a sphere, as $\mathrm{d}S$ is no longer the area of an infinitesimal rectangle.

Certainly, we should expect dS to depend on the sphere's radius—it should be proportional to a^2, as this is how the sphere's area scales with its radius. But dS may also depend on the co-ordinates θ and φ. The shaded cap shown in Figure 36 is the region given by $0 \leqslant \theta \leqslant \alpha$. If we increase α to $\alpha + d\theta$, then the shaded area grows at different rates depending on the value of α, even when using the same increment $d\theta$. If we're near either pole, then the area barely increases, but near the equator it increases the most, with the extra area making a band around the sphere at its 'fattest'. By comparison, the symmetry that the sphere has about its north–south pole axis implies that an increment in φ has the same effect, irrespective of where we are on the sphere.

The correct formula for an infinitesimal element of surface area is

$$dS = a^2 \sin(\theta) d\theta \, d\varphi.$$

Note that when we are at the poles—that is, $\theta = 0$ or $\theta = \pi$—then $\sin(\theta) = 0$, and when we are at the equator—that is, $\theta = \pi/2$—then $\sin(\theta) = 1$, reflecting the different increments in surface area that the same change in θ can produce. We can now determine the surface area of the shaded cap as

$$\begin{aligned} &\int_{\theta=0}^{\theta=\alpha} \int_{\varphi=0}^{\varphi=2\pi} a^2 \sin(\theta) d\varphi \, d\theta \\ &= a^2 \Big(-\cos(\alpha) - (-\cos(0)) \Big) (2\pi - 0) \\ &= 2\pi a^2 (1 - \cos(\alpha)) \end{aligned}$$

recalling that $-\cos(\theta)$ is an antiderivative of $\sin(\theta)$, and using the fundamental theorem. At the latitude of the south pole, the cap becomes the whole sphere; there $\alpha = \pi$ and $\cos(\alpha) = -1$ so that we obtain $4\pi a^2$ as expected.

The above calculation leads naturally to a discussion of **solid angle**. In the plane, the angle between two half-lines equals the arc length of the unit (= radius 1) circle between the half-lines. This is

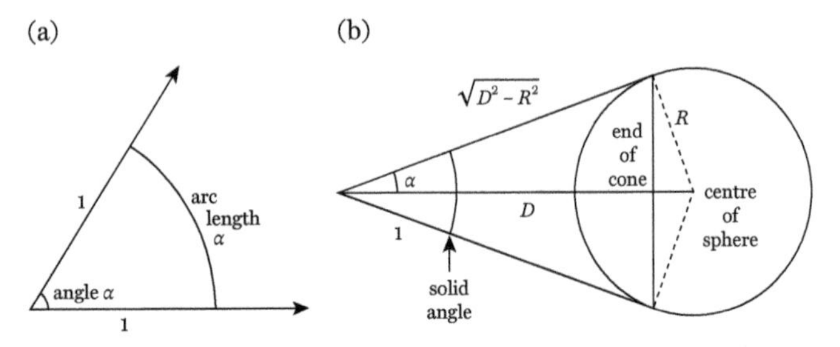

37(a). Plane angle. 37(b). Solid angle in a cone.

provided we are using radians to measure angles (Figure 37(a)). In a similar fashion, we can define the solid angle at the apex of a cone as the surface area of the unit sphere that the cone cuts out (Figure 37(b)). If the semi-angle in the cone is α, then the previous calculation shows that the solid angle is equal to $2\pi(1 - \cos(\alpha))$.

The unit of solid angle is the **steradian**. In the same way that the unit circle having length 2π means that there are 2π radians in a whole angle, the unit sphere having area 4π means that there are 4π steradians in a whole solid angle.

We can now work out the solid angle when looking at the Sun or Moon. Each can be modelled as a sphere, with radius R at distance D from the observer (Figure 37(b)). We see that

$$\cos(\alpha) = \frac{\sqrt{D^2 - R^2}}{D} = \sqrt{1 - \left(\frac{R}{D}\right)^2}.$$

The approximate values of R and D for the Sun and Moon are:

$$R_{\text{sun}} = 7 \times 10^5 \text{km}, \quad D_{\text{sun}} = 1.5 \times 10^8 \text{km}, \quad \frac{R_{\text{sun}}}{D_{\text{sun}}} = 4.67 \times 10^{-3};$$
$$R_{\text{moon}} = 1.8 \times 10^3 \text{km}, \quad D_{\text{moon}} = 3.8 \times 10^5 \text{km}, \quad \frac{R_{\text{moon}}}{D_{\text{moon}}} = 4.73 \times 10^{-3}.$$

Consequently, the Sun and Moon occupy much the same portion of the sky, despite being considerably different in their actual sizes.

It's somewhat more complicated to calculate the solid angle made by a more irregular object. If you look around you at an object, and envisage a sphere of unit radius centred on one of your eyes, then the solid angle made is how much of that sphere is taken up by your view of the object. The surface of the object you're looking at is made up of infinitesimal elements $\mathrm{d}S$ of surface area; however, the same amount of surface area will obstruct differing amounts of your vision depending on how far away the object is and how obliquely you are looking at it.

If you were looking at part of a sphere with area S, all of which was distance r away, then it would make a solid angle S/r^2. The scale factor of $1/r^2$ represents how your view of an object diminishes as it recedes; equally, r^2 represents how a sphere's area grows as its radius r increases. For most objects, though, different parts of an object are at different distances from your eye, so we would need to use an individual value of r for each part. In addition, the same surface area, at a certain distance away, can obstruct more or less of your view depending how full-on it appears to you. A sheet of paper might be almost invisible if close to being on its side, in contrast to when looked at fully. Putting all this together, the solid angle of an irregular object is defined to be the integral over the viewed surface of the obstructive component of each scaled element of area $\mathrm{d}S/r^2$.

Solid angle is an example of a **flux integral**. Flux is an important notion in fluid dynamics, thermodynamics, and electromagnetism, and is a measure of the rate at which a quantity passes through a surface.

For example, suppose that a fluid has constant velocity v when moving through a pipe with cross-sectional area A. Then the amount of fluid passing the cross-section, or flux, is vA per second. Here, the fluid is moving perpendicularly to the cross-section, but

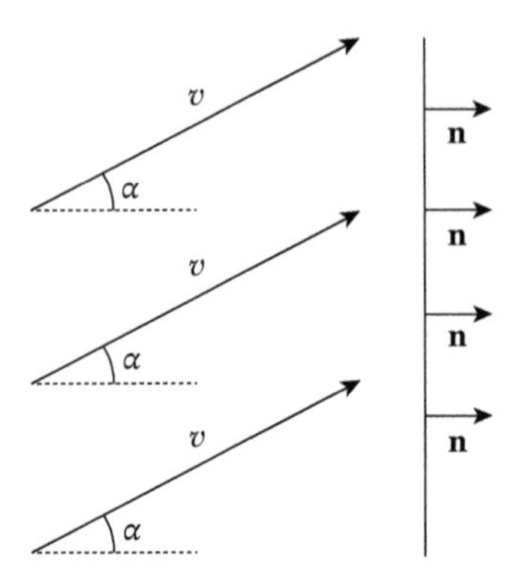

38. Flux.

if the fluid moves obliquely to the boundary (Figure 38), then the flux crossing an area A equals $vA\cos(\alpha)$. Note that when $\alpha = 0$, the flux equals vA, as the fluid is moving perpendicular to the boundary, while when $\alpha = \pi/2$, the flux equals 0, as the fluid is moving parallel to the boundary, rather than through it.

This expression $v\cos(\alpha)$ is commonly denoted by $\mathbf{v} \cdot \mathbf{n}$, where $\mathbf{n}$ is the unit vector perpendicular to the surface and $\mathbf{v}$ is the velocity vector of the fluid. (Some readers may recognize this as a scalar product.) More generally, the velocity $\mathbf{v}$ of a fluid is not constant and nor is the perpendicular unit vector $\mathbf{n}$ to a surface constant, so the flux across a general surface X equals a sum of the flux across infinitesimal elements $\mathrm{d}S$ of surface—that is, the flux equals

$$\iint_X \mathbf{v} \cdot \mathbf{n}\, \mathrm{d}S.$$

Line integrals and work

Much of what follows in this chapter first interested applied mathematicians and physicists. The nature of these integrals gives

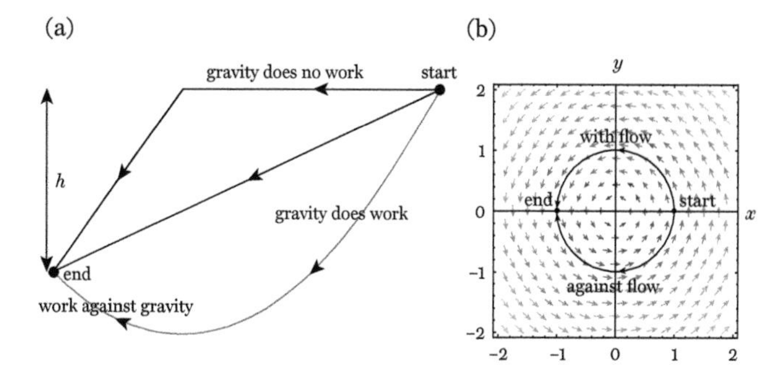

39(a). Work done by gravity. 39(b). A whirlpool $\mathbf{v}(x, y) = (-y, x)$.

a sense of their physical origins; for example, line integrals are perhaps best introduced by thinking of the work done by a force.

Gravity acts around us in a (relatively) uniform, constant way—we model acceleration due to gravity as the vector $(0, 0, -g)$. That is, gravity acts downwards with constant strength g, which is approximately 9.81ms^{-2}. By Newton's second law, the gravitational force on an object of mass m equals $\mathbf{F} = (0, 0, -mg)$.

If we lower that object by a height of h, then gravity has done **work** and the object has lost *gravitational potential energy* in the amount of mgh. An important point here is that the energy lost equals mgh, irrespective of which path is taken between the start and end of the journey (Figure 39(a)). By contrast, we can look at a vector flow representing a whirlpool at the origin (Figure 39(b)). Here, it's clear that if you swam between the start and end points labelled, different amounts of work would be done depending on whether you swam with the flow (anticlockwise) or against it (clockwise).

If an object were moved by a constant force F in a straight line distance d, the work done would equal Fd. More generally, the force $\mathbf{F}$ would not be constant, and nor would the path taken be a straight line. For example (Figure 39(a)), while the top path is horizontal no work is being done, because gravity has no component in the direction of travel; if the path is vertical the full force of gravity is acting, and at other points some component of gravity is acting.

For a general motion, if we move an infinitesimal amount,

$$\mathbf{dr} = (\mathrm{d}x, \mathrm{d}y, \mathrm{d}z),$$

in three-dimensional space and in the presence of a force $\mathbf{F} = (F_x, F_y, F_z)$, then the infinitesimal work done by the force is denoted $\mathbf{F} \bullet \mathbf{dr}$ and equals

$$\mathbf{F} \bullet \mathbf{dr} = F_x \mathrm{d}x + F_y \mathrm{d}y + F_z \mathrm{d}z.$$

In this expression, the horizontal component of the movement $(\mathrm{d}x, 0, 0)$ picks out the horizontal component of the force $(F_x, 0, 0)$, and likewise for the other co-ordinates. At the points where the path is horizontal, $\mathrm{d}z = 0$, and so $\mathbf{F} \bullet \mathbf{dr} = 0$ when $\mathbf{F}$ is the gravitational force, as $F_x = F_y = 0$ as well (Figure 39(a)). The total amount of work done, as we move along a curve C in a certain direction, is denoted by

$$\int_C \mathbf{F} \bullet \mathbf{dr},$$

and equals the sum of all the infinitesimal work contributions as we make the journey along the curve C. If we move along C in the opposite direction, the work done changes signs.

For gravity $\mathbf{F} = (0, 0, -mg)$ we have $\mathbf{F} \bullet \mathbf{dr} = -mg\mathrm{d}z$ and $-mg$ has antiderivative $-mgz$. So, for any curve C starting at a point P which is height h above the finishing point Q,

$$\int_C \mathbf{F} \bullet \mathbf{dr} = -\int_P^Q mg \ \mathrm{d}z = -mg(z(Q) - z(P)) = -mg(-h) = mgh,$$

by the fundamental theorem. What makes the gravitational force **F** special in this regard is that **F** is **conservative**, which means that $\mathbf{F} = \text{grad}(f)$ is the gradient vector of a function f called a **potential**. When $\mathbf{F} = \text{grad}(f)$ is conservative, we have

$$\int_P^Q \text{grad}(f) \cdot \mathbf{dr} = f(Q) - f(P),$$

which is essentially just another version of the fundamental theorem. It says that the work done by a conservative force along a curve C is just the change in potential as we move from P, the start of C, to its end Q. For gravity, $\mathbf{F} = \text{grad}(f)$ where $f = -mgz$.

When the path is a loop, so that P and Q are the same point, no work is done, which is why such fields are termed conservative. Most fields aren't conservative: for example, the 'whirlpool' seen previously (Figure 39(b)) isn't, because different amounts of work are done going around the two semi-circles even though they have the same starting points and endpoints.

Stokes' theorem and the divergence theorem

A vector field **v** in two or three dimensions can represent the velocity of a fluid. For such flows, there are two further important properties: their **curl** and **divergence**.

The divergence $\text{div}(\mathbf{v})$ of a vector field $\mathbf{v} = (v_x, v_y, v_z)$ in three dimensions is given by

$$\text{div}(\mathbf{v}) = \frac{\partial v_x}{\partial x} + \frac{\partial v_y}{\partial y} + \frac{\partial v_z}{\partial z}.$$

(In two dimensions the formula involves just the first two terms on the right-hand side.) The physical meaning of divergence relates to the expansion or contraction of the fluid's flow. The vector field $\mathbf{v} = (x, y)$, shown (Figure 40), has divergence $\operatorname{div}(\mathbf{v}) = \partial x/\partial x + \partial y/\partial y = 1 + 1 = 2$. Also drawn is the progress of a disc of fluid as it expands with time t while following the flow $\mathbf{v}$. If $A(t)$ is the area of the disc at time t, we have

$$\frac{\mathrm{d}A}{\mathrm{d}t} = 2A.$$

The number 2 in this differential equation is the divergence of the flow. More generally, the divergence isn't constant but measures the rate of local expansion (if positive) or contraction (if negative) of the fluid at a particular point and time. The condition that $\operatorname{div}(\mathbf{v}) = 0$ means that a fluid is **incompressible**, as is the case with

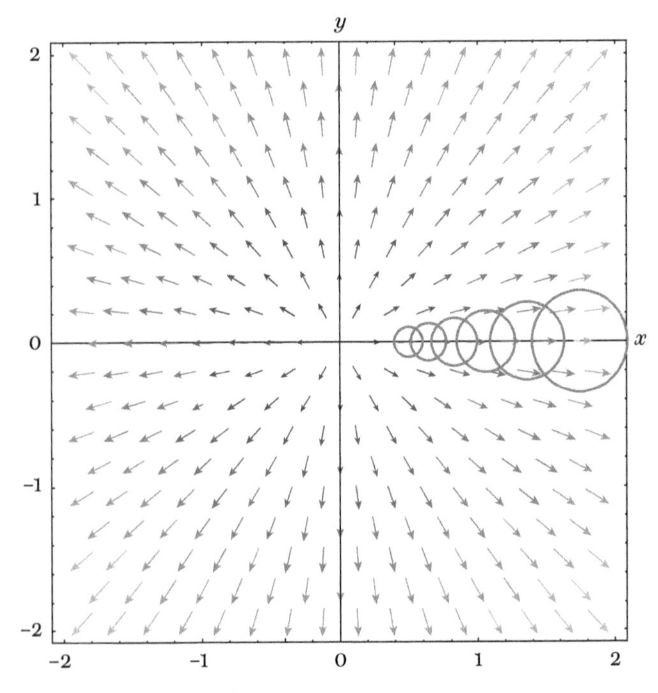

40. $\mathbf{v}(x, y) = (x, y)$ **with div** $\mathbf{v} = 2$.

the rotating flow seen previously (Figure 39(b)). Incompressible fluids are used in hydraulic machinery; the pressure applied to an incompressible fluid is transmitted with very little loss and can be directed along pipes and around corners in a versatile way.

The **divergence theorem** is a higher-dimensional version of the fundamental theorem of calculus. It states that

$$\iiint_R \operatorname{div}(\mathbf{v})\mathrm{d}V = \iint_{\partial R} \mathbf{v} \bullet \mathbf{n} \ \mathrm{d}S,$$

where:

- $\mathbf{v}$ represents a fluid's velocity;
- R is a bounded region of three-dimensional space, such as a solid sphere or cube;
- ∂R is the surface of R, i.e. the spherical shell or six faces of the cube for the above examples;
- $\mathbf{v} \bullet \mathbf{n}$ is the rate of fluid leaving R across the boundary ∂R per unit area.

Given an infinitesimal volume of fluid $\mathrm{d}V$, $\operatorname{div}(\mathbf{v})\mathrm{d}V$ is the rate at which it is expanding or contracting, so the triple integral on the left of the divergence theorem is the sum of all these expansions and contractions inside the region R. On the right-hand side is the total flux of the fluid across the boundary of R. So, the divergence theorem, physically interpreted, states that the overall rate of expansion and contraction inside a region equals the rate of fluid leaking out and in across the boundary.

A physical interpretation of curl is comparatively subtle, and the detailed formula for curl is also rather involved, so is left to the Appendix. As an example, though, imagine the xy-plane rotating about the origin (Figure 41). If ω is the **angular speed**—meaning that the rotation happens at a rate of ω radians per second—the velocity of the point (x, y) is given by $\mathbf{v} = (-\omega y, \omega x)$ and

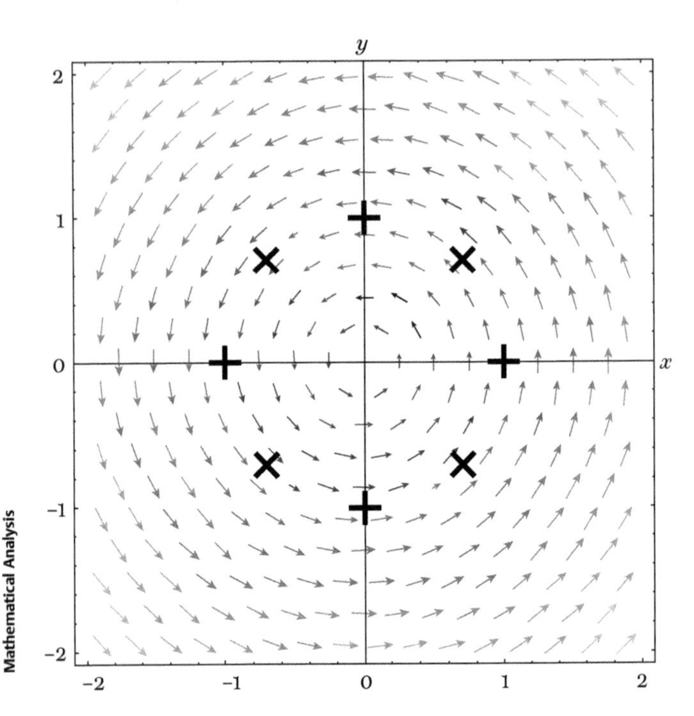

41. Rotation around the origin.

$\text{curl}(\mathbf{v}) = (0, 0, 2\omega)$. So here, the magnitude of curl is twice the angular speed, and its direction, $(0, 0, 1)$, is parallel to the axis of the rotation, in this case the z-axis.

For a general fluid flow, $\text{curl}(\mathbf{v})$ is a measure of the flow's **vorticity**, and it's important to recognize that this is a measure of local behaviour. In the previous example the whole plane is rotating with the same angular velocity. More generally, for a small element of fluid moving in a flow $\mathbf{v}$, the direction of $\text{curl}(\mathbf{v})$ is parallel to the element's axis of rotation and the magnitude of $\text{curl}(\mathbf{v})$ is twice the angular speed of the element's rotation. A fluid flow with zero curl is called **irrotational**.

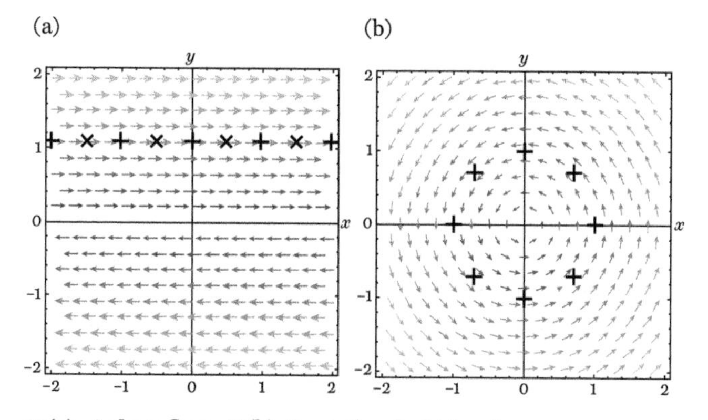

42(a). A shear flow. 42(b). Irrotational whirlpool.

Note that all points in Figure 41 are rotating with angular speed ω; this can be appreciated by noting how the small crosses '+' are each rotating. The turning of these crosses is what's important, not the fact that points are moving globally in circles; the turning crosses represent how the flow is spinning locally. The diagram on the left (Figure 42(a)) shows a *shear flow*. Globally, particles move in straight lines, but they themselves are spinning while doing so—as exhibited by the spinning crosses—and the curl is non-zero. By contrast, the diagram on the right (Figure 42(b)) illustrates a flow with zero curl but with particles moving globally in circles, without spinning while doing so, as exhibited by the crosses' behaviour.

Again, there is a higher-dimensional version of the fundamental theorem which involves curl. This is called **Stokes' theorem**, which states that

$$\iint_R \operatorname{curl}(\mathbf{v}) \cdot \mathbf{n} \, dS = \int_{\partial R} \mathbf{v} \cdot d\mathbf{r},$$

where:

- $\mathbf{v}$ represents a fluid's velocity;
- R is a surface situated in three dimensions, such as a hemisphere;

- ∂R is the bounding curve of R, so a circle when R is a hemisphere;
- $\mathbf{n}$ is a unit vector, perpendicular to the surface ∂R.

A physical interpretation of Stokes' theorem is possible, but again is subtle. Curl measures twice the local spin of a vector field. We might visualize the curl of the field as spinning cogs on the surface R intermeshing with one another. The flux integral of the left-hand side of Stokes' theorem is the sum of all the cogs' contributions. No internal cog is driving the others; they are just simultaneously moving together, and no work is being done. However, for the cogs on the boundary, half their contact is with internal cogs but they have no contact on the other side of the boundary. If there were a caterpillar track on the boundary, then the external cogs would be doing work driving that track around, and this is what is measured by the work integral on the right-hand side of Stokes' theorem. This is captured somewhat simplistically in the diagram below (Figure 43), where we see all the contributions cancel on the internal edges.

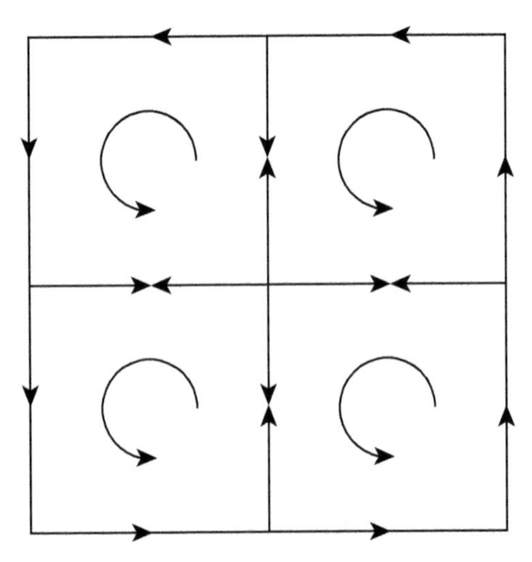

43. Interpreting Stokes' theorem physically.

Stokes' theorem is named after the Irish physicist George Stokes, though it should more properly be named after William Thomson (later Lord Kelvin), who wrote to Stokes with details of the theorem in 1850. Stokes' name became attached to the theorem after he posed it as a University of Cambridge examination question in 1854. The special case of R being a region of the xy-plane is known as **Green's theorem**, named after George Green, who earlier stated his theorem in *An Essay on the Application of Mathematical Analysis to the Theories of Electricity and Magnetism* in 1828 (see Appendix). Special cases of the divergence theorem had been known to Lagrange and Gauss, but it was Mikhail Ostrogradsky who gave the first complete proof in 1826 while studying heat flow.

A final point of note is that the curl of a conservative vector field is zero. This is a relatively simple matter of algebra (see Appendix). More interesting is the converse question: if curl($\mathbf{v}$) is zero, is $\mathbf{v}$ conservative? Surprisingly, the answer is positive or negative depending on the *shape* of the region that $\mathbf{v}$ is defined on.

The answer is positive if $\mathbf{v}$ is defined on the whole xy-plane or whole three-dimensional xyz-space. However, the flow

$$\mathbf{v}(x,y) = \left(\frac{-y}{x^2+y^2}, \frac{x}{x^2+y^2}\right)$$

(Figure 42(b)) is an example of a vector field with zero curl which is not conservative—that is, no potential exists. Note here that $\mathbf{v}$ is not defined at the origin where $x = y = 0$; so $\mathbf{v}$ is defined on a punctured plane, a plane missing a point, and a region of crucially different shape which is not *simply connected* (informally, it has holes in it). For simply connected regions, the converse *is* true. We are now straying into the field of *differential topology*, which is where the study of these theorems properly lies.

The fundamental theorem, Stokes' theorem, and the divergence theorem are one-, two-, and three-dimensional versions of the same theorem. This was made explicit by Élie Cartan, who proved the **generalized Stokes' theorem** in 1945. This modern form of Stokes' theorem has an incredibly succinct statement, namely

$$\int_M \mathrm{d}\omega = \int_{\partial M} \omega,$$

though it would take considerable effort to explain it in full. Here M is a *manifold*—which is a higher-dimensional equivalent of a curve or surface—which has boundary ∂M.

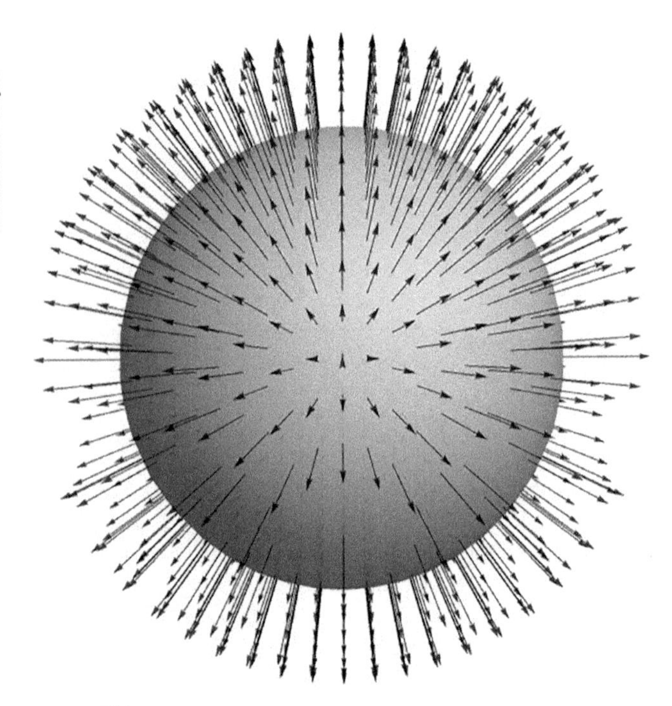

44. grad(g) on g = 1.

There is a large amount of theory relating to analysis on such spaces. One aspect relates to constrained optimization problems; for example, what is the maximal value that $f(x, y, z) = 2x + 2y + z$ takes when $g(x, y, z) = x^2 + y^2 + z^2 = 1$? The constraint $g = 1$ means that (x, y, z) lies on a sphere, so we are asking what is the largest value that f attains on that sphere. The greatest value is 3 which is achieved when $(x, y, z) = (2/3, 2/3, 1/3)$. Note at this point that $\text{grad}(f) = (2, 2, 1)$ and

$$\text{grad}(g) = (2x, 2y, 2z) = \left(\frac{4}{3}, \frac{4}{3}, \frac{2}{3}\right) = \frac{2}{3}\text{grad}(f).$$

The vectors grad(f) and grad(g) are parallel; this is not a coincidence, but rather what the theory of *Lagrangian multipliers* states will occur at such a maximum or minimum. This is because the gradient vector grad(g) is everywhere perpendicular to the surface $g = 1$ (Figure 44), whilst grad(f) is perpendicular to the surface at a maximum or minimum of f (in a result akin to Fermat's theorem).

The huge field of *differential geometry* is essentially the application of multivariable calculus to the geometry of curves, surfaces, and, more generally, manifolds. The *topology* of a space, essentially its shape, can have important global implications for analysis on that space. See, for example, *Topology: A Very Short Introduction*.

Chapter 6
I'll name that tune in . . .

The wave equation

An early mathematical model employing calculus was the *wave equation*, derived by the French mathematician and music theorist Jean le Rond D'Alembert. This equation governs how a taut string—such as a guitar or piano string—makes small vibrations, but it is also important in the study of acoustics more generally, electromagnetism (including light), and fluid dynamics.

There are various ways to make a string vibrate. In a harpsichord, the string is plucked and released; in a piano, a string at rest is struck by a 'hammer'. How can such vibrations be described mathematically? At a particular instant of time t, a vibrating string will make a certain shape, which we can represent as the graph of a function $y(x)$. Here x is the co-ordinate of a point along the string, and the value $y(x)$ is the sideways—or transverse—displacement of that point, measured from the equilibrium position $y = 0$. The notion of displacement here is similar to that of distance, but because the string might be above or below the horizontal equilibrium, y can be positive or negative. Combining all this means that we need a function, $y(x, t)$, of two variables, x and t, to fully describe the string's position.

This will be sufficient to describe any **transverse** vibration of the string, meaning that a point of the string with horizontal

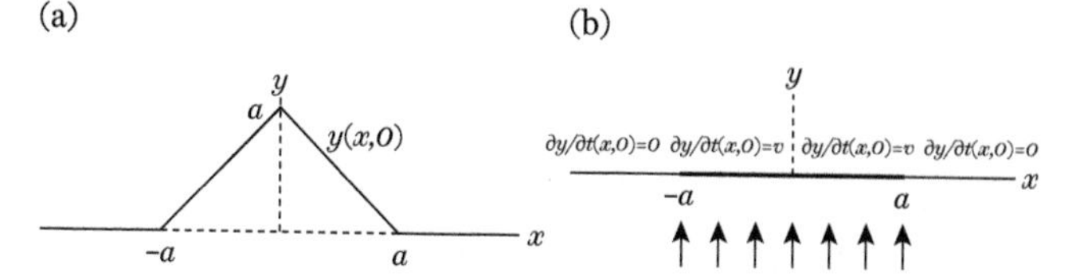

45(a). A plucked string initially. 45(b). A struck string initially.

co-ordinate x maintains that co-ordinate, moving only in an 'up-and-down' fashion. This is a reasonable assumption for a taut string, making small vibrations, but would not apply if we had a loose string. The transverse velocity of the string is given by the partial derivative $\frac{\partial y}{\partial t}$; this is the velocity at which a particular point of the string is moving up or down. We will assume that the string has a constant tension T throughout and a uniform density ρ.

The initial position $y(x, 0)$ of a harpsichord string at time $t = 0$ is graphed (Figure 45(a)). Here $\frac{\partial y}{\partial t}(x, 0) = 0$, signifying that the string is initially at rest. By contrast, a piano string begins in the equilibrium position $y(x, 0) = 0$ (Figure 45(b)), but the string is initially struck by a hammer, so a part of the string, $-a < x < a$, has an initial transverse velocity $\frac{\partial y}{\partial t}(x, 0) = v$.

In 1747 D'Alembert derived what is now known as the **wave equation**, which states that

$$T\frac{\partial^2 y}{\partial x^2} = \rho\frac{\partial^2 y}{\partial t^2}.$$

Any small transverse vibration of the string is a solution $y(x, t)$ of the wave equation.

The wave equation models how these two initial states evolve with time (Figures 46(a), 46(b)). For a plucked string, half the initial pluck moves to the right and half moves to the left. Both halves move with speed $c = \sqrt{T/\rho}$. For a struck string, a central portion of the

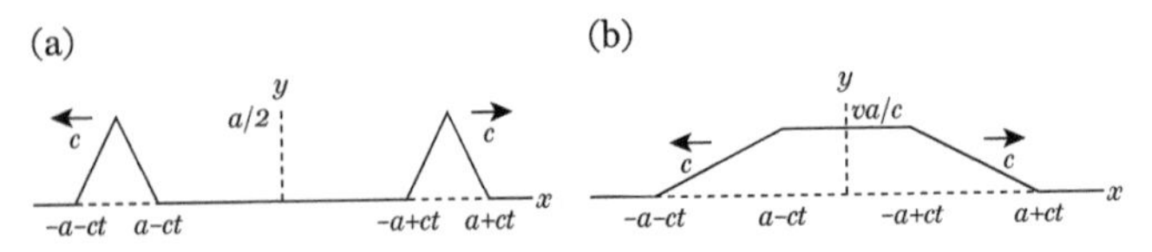

46(a). A plucked string propagates. 46(b). A struck string propagates.

string eventually moves through a distance va/c and then stops, with the front and rear of the wave moving right and left, again with speed c.

The wave equation is a partial differential equation, as discussed in Chapter 5; further, it is a second-order equation, so we might expect the general solution to include two arbitrary functions. D'Alembert solved the wave equation for an infinite string, showing

$$y(x,t) = f(x - ct) + g(x + ct),$$

where f and g are arbitrary functions. The quantity c is the speed at which a wave propagates along the wire. So, if the tension T is greater, waves propagate faster and, if the density ρ is greater, the waves move more slowly. As all small vibrations satisfy the wave equation, we need further information, such as the initial position and initial motion of the string, to work out specifically what f and g are.

A solution of the form $y(x,t) = f(x - ct)$ consists of the graph of $y = f(x)$ moving right along the string at speed c; to appreciate this, note that $x + ct$ is a point moving right with speed c, starting at x, and $f(x - ct)$ takes the same value at $(x + ct, t)$ as it did at $(x, 0)$. Similarly a solution of the form $y(x,t) = g(x + ct)$ consists of the graph of $y = g(x)$ moving left along the string at speed c. The general solution therefore comprises two waves moving left and right with speed c.

Historically, however, there was still an issue with what should be understood by an 'arbitrary function'. The initial starting position of $y(x, 0)$ for the plucked string is physically reasonable (Figure 45(a)), but $y(x, 0)$ wouldn't have met Euler's definition of a function being described by a single analytic expression. 18th-century mathematicians were aware of this problem, though it went unresolved at the time.

Derivation of the wave equation

Recall that a mathematical model is a description of how the real world behaves. It is necessarily something of an approximation, and reasonable, though idealized, simplifications are made so that the mathematics involved is not overcomplicated. What follows is a brief description of the wave equation's derivation. It is a somewhat technical and physical argument, so you may prefer to move on to the next section, but it is included here because it is typical of how the methods of calculus get applied to real-world models. Such models form the most important application of calculus.

Models often begin by focusing on a small part of an experiment, and as we let that part become ever smaller, limits and derivatives arise in our calculations. So we begin with a very small piece of the string, say of length h, which runs from x to $x + h$. The mass of this small piece of string is ρh. Depicted in Figure 47 is this piece lying flat, below its later position.

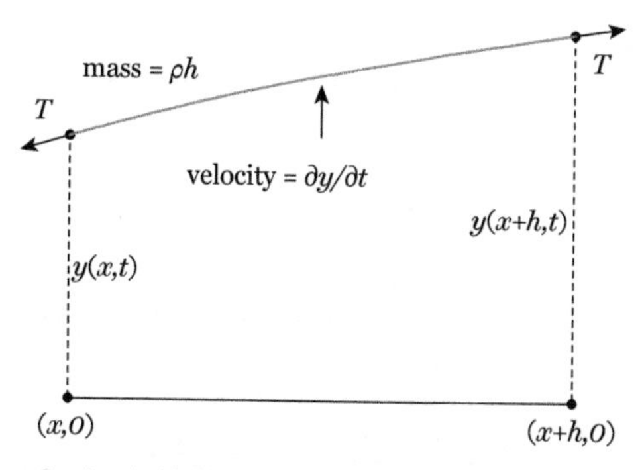

47. Piece of string initially and then vibrating.

The figure shows the forces acting on this piece of string, namely tension pulling the string to the right and to the left. We (reasonably) ignore gravitational effects or air resistance, as these forces are negligible compared with the tension T in the string. Because the piece of string moves only up and down—so not sideways—the component of the tension pulling the string to the right must cancel out the component pulling it left. It's the difference between the up and down components of the tension at the two ends that makes the string move up and down.

The gradient of the string at any point is $\frac{\partial y}{\partial x}$; from this it follows that the upward component of the tension on the right is $T\frac{\partial y}{\partial x}(x+h,t)$ and the downward component of the tension on the left is $T\frac{\partial y}{\partial x}(x,t)$. The string's mass is ρh and the upward acceleration of the piece is $\partial^2 y/\partial t^2$. Putting all these quantities into Newton's second law, that force = mass × acceleration, gives

$$T\frac{\partial y}{\partial x}(x+h,t) - T\frac{\partial y}{\partial x}(x,t) = (\rho h)\frac{\partial^2 y}{\partial t^2}.$$

If we divide both sides of the equation by h and take the limit as h becomes small, then we arrive at

$$T\frac{\partial^2 y}{\partial x^2} = \rho\frac{\partial^2 y}{\partial t^2};$$

this is the wave equation. For those interested in finding out more about mathematical models, see *Applied Mathematics: A Very Short Introduction*.

Boundary value problems

Clearly a guitar or piano string is not infinite and, besides being of finite length L, both its ends are secured. So the displacement $y(x,t)$ must satisfy the wave equation

$$c^2\frac{\partial^2 y}{\partial x^2} = \frac{\partial^2 y}{\partial t^2} \qquad \text{where } c^2 = \frac{T}{\rho}$$

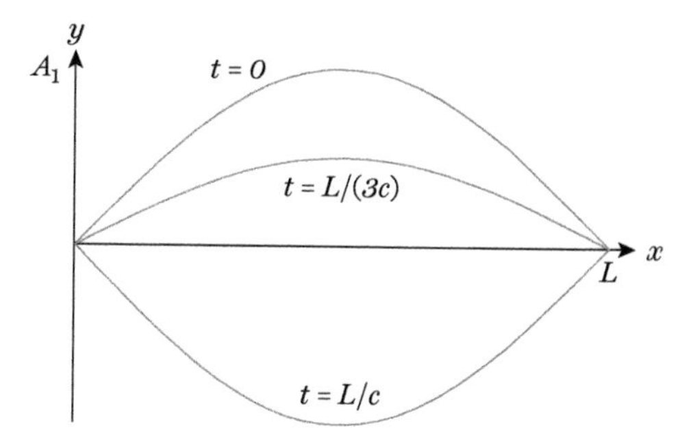

48. First harmonic at $t = 0$, $L/(3c)$ and $t = L/c$.

and *also* the boundary conditions

$$y(0,t) = 0, \quad y(L,t) = 0.$$

These boundary conditions say that the start of the string $x = 0$ and the end of the string $x = L$ are secured to the x-axis (where $y = 0$) at all times. One possible mode of vibration for such a string, which is a solution to the wave equation and its boundary conditions, is

$$y_1(x,t) = A_1 \sin\left(\frac{\pi x}{L}\right)\cos\left(\frac{\pi ct}{L}\right), \quad \text{for } 0 \leqslant x \leqslant L.$$

This manner of vibration is called the string's **first harmonic**, here sketched at three different times (Figure 48). When $t = 0$, the string is at its highest and the very highest the string achieves, A_1, is called the *amplitude*; as time passes, the string moves down to be at its lowest when $t = L/c$, the mirror image of its original position; it then moves up, returning to its original position when $t = 2L/c$, which is the *period* of a single oscillation.

The second harmonic (Figure 49(a)) is given by

$$y_2(x,t) = A_2 \sin\left(\frac{2\pi x}{L}\right)\cos\left(\frac{2\pi ct}{L}\right), \quad 0 \leqslant x \leqslant L.$$

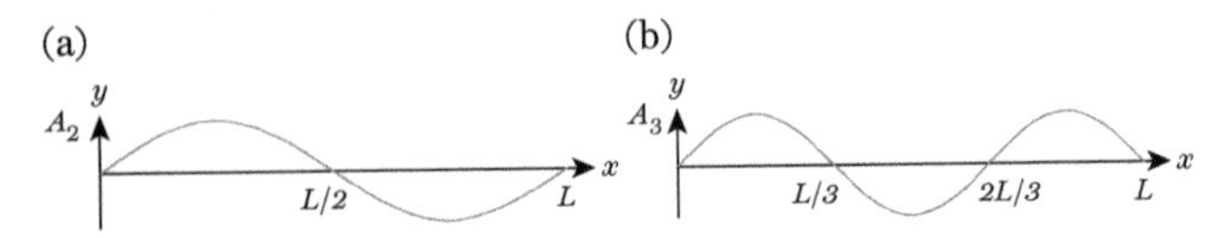

49(a). Second harmonic. 49(b). Third harmonic.

Its period is half that of the first harmonic, and its pitch is an *octave* higher than the first harmonic.

More generally, for $n = 1, 2, 3, \ldots$, the string can vibrate as follows:

$$y_n(x,t) = A_n \sin\left(\frac{n\pi x}{L}\right)\cos\left(\frac{n\pi ct}{L}\right), \quad 0 \leqslant x \leqslant L,$$

which is known as the nth harmonic. This oscillation has a period that is n times shorter than the first harmonic. The third harmonic (Figure 49(b)) has a pitch that is a *fifth* above that of the second harmonic.

These harmonics are very special ways in which a string can vibrate. In general, a string will vibrate in a manner that involves many, indeed infinitely many, of the harmonics. Finding out how much of each harmonic contributes to a particular vibration is the study of *Fourier analysis*, named after the French mathematician Joseph Fourier.

Fourier analysis

In general, the vibrations of a string $y(x,t)$ with fixed ends are not single harmonics but rather are a combination of different harmonics. So $y(x,t)$ can be written as an infinite sum,

$$y(x,t) = y_1(x,t) + y_2(x,t) + y_3(x,t) + \ldots,$$

of harmonics. The function

$$y_n(x,t) = A_n \sin\left(\frac{n\pi x}{L}\right)\cos\left(\frac{n\pi ct}{L}\right)$$

is an nth harmonic, and our task is to find the amplitude A_n of each component for a general mode of vibration $y(x,t)$. Fourier showed that if the string is initially at rest in the shape of the graph $y = f(x)$, then the amplitude A_n equals the integral

$$A_n = \frac{2}{L}\int_0^L f(x)\sin\left(\frac{n\pi x}{L}\right)\mathrm{d}x$$

(as derived in the Appendix). Using these values of A_n in the formula for $y_n(x,t)$ and substituting those expressions for $y_n(x,t)$ in the infinite sum for $y(x,t)$, we obtain the solution to the wave equation for a vibrating string starting from rest. The expression of $y(x,t)$ as a combination of harmonics is called its **Fourier series.** The values A_n are called **Fourier coefficients.**

As a first example of a Fourier series, we will set $L = \pi$ (which simplifies the calculations a little) and consider a plucked string at rest at time $t = 0$ (Figure 50(a)).

The integrals A_n for the plucked string can be evaluated, and the plucked string's Fourier series turns out to equal

$$\frac{4}{\pi}\left(\sin(x) - \frac{1}{3^2}\sin(3x) + \frac{1}{5^2}\sin(5x) - \frac{1}{7^2}\sin(7x) + \cdots\right).$$

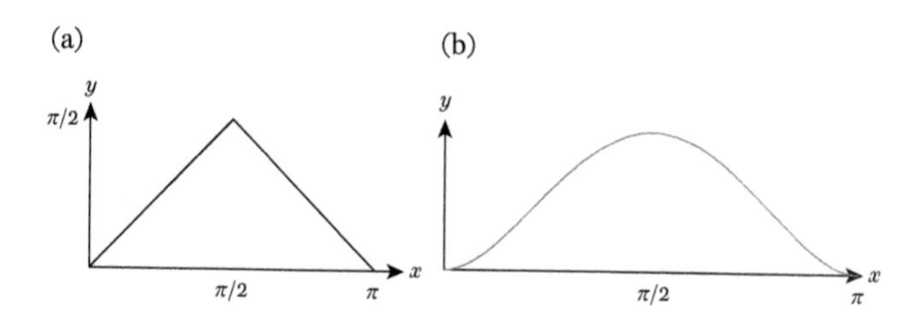

50(a). Plucked string. 50(b). $f(x) = x^2(x-\pi)^2$**.**

If we set $x = \pi/2$, the above sum converges to $f(\pi/2) = \pi/2$, and, rearranging, we would deduce

$$\frac{\pi^2}{8} = 1 + \frac{1}{3^2} + \frac{1}{5^2} + \frac{1}{7^2} + \frac{1}{9^2} + \cdots,$$

which is correct. In fact, the above Fourier series converges to the plucked string function for all x in the range $0 \leqslant x \leqslant \pi$. More generally, the convergence of Fourier series turns out to be quite a subtle matter.

The impact of Fourier's work

Fourier analysis was initially developed to model heat flow in solid bodies; Fourier's seminal work *The Analytical Theory of Heat* was published in 1822, though his initial results date back to 1807. He derived the *heat equation* which models heat flow in solid bodies; this is another partial differential equation which can be solved using Fourier analysis in a manner similar to how the wave equation was solved. Since then, Fourier analysis—including the related Fourier transform—has found many applications in mathematics and science, in the study of partial differential equations and in acoustics, optics, NMR (nuclear magnetic resonance), and signal and image processing.

Fourier was not the first to employ trigonometric series to solve partial differential equations—as early as 1753 Daniel Bernoulli had argued for solving the wave equation with such series. Fourier, though, determined formulae for the Fourier coefficients and recognized the generality of his methods. In fact, he was initially overconfident in their use, claiming (incorrectly) that any function—continuous or discontinuous—could be represented by an infinite trigonometric series.

The convergence—or otherwise—of Fourier series then became an important question in analysis, with much progress being made in

1829 by Gustav Dirichlet in a significant memoir. If a function is differentiable, then its Fourier series converges to the function. Fourier series even handle jump discontinuities in such functions and converge to the average of the function either side of the discontinuity. However, in 1873, a continuous function was found with a Fourier series that did not converge to the function.

The convergence of a Fourier series is intimately linked with the function's differentiability. The **Riemann–Lebesgue lemma** shows that the Fourier coefficients necessarily converge to zero as n increases, but the speed of that convergence depends on the differentiability of the function. The coefficients of the plucked string (Figure 50(a)) decrease like $1/n^2$. Once calculated, the Fourier coefficients of $x^2(x-\pi)^2$ decrease more quickly, like $1/n^3$ (Figure 50(b)). This is because the plucked string has points, such as the apex, where it is not differentiable, whereas $x^2(x-\pi)^2$ is differentiable everywhere. This pattern continues, with coefficients decreasing ever faster for functions that are twice differentiable, and so on. A further aspect is *Gibbs' phenomenon*: Fourier series converge poorly at discontinuities of a function or its derivatives. For example, the plucked string and its approximation using the first four non-zero terms of its Fourier series differ most at the apex where the derivative is discontinuous (Figure 51).

Beyond their applications, the investigation into their convergence would provide enormous impetus for mathematicians to think

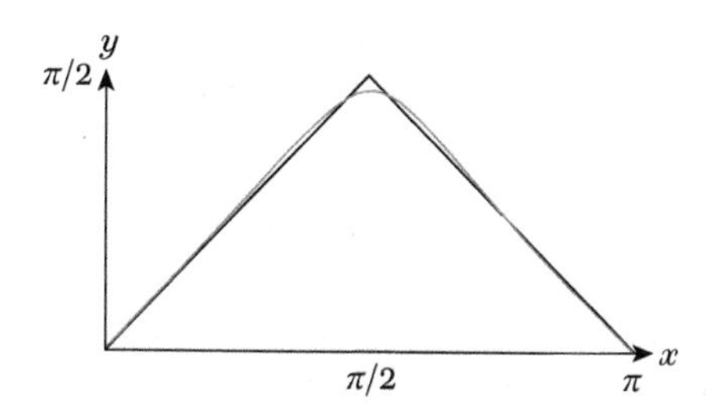

51. Gibbs' phenomenon.

about functions more broadly—such as in Riemann's theory of integration—and make clearer the need for rigour in analysis. Cantor's work on different infinities (Chapter 1) began with his study of Fourier series. Fourier analysis also raised the question: what is special about trigonometric functions that makes Fourier series so powerful? Put another way, what other infinite sums of functions should mathematicians be considering?

Can you hear the shape of a drum?

A natural generalization of the wave equation governs how two-dimensional membranes, such as a circular drum or a rectangular membrane, can vibrate. For a rectangular membrane with sides of length L_1 and L_2 and a fixed boundary, the harmonics are given by

$$z(x,y,t) = A_{m,n} \sin\left(\frac{m\pi x}{L_1}\right)\sin\left(\frac{n\pi y}{L_2}\right)\cos\left(\sqrt{\frac{m^2}{{L_1}^2}+\frac{n^2}{{L_2}^2}}\pi ct\right),$$

where m and n are positive integers. The period of the above harmonic is

$$\frac{2/c}{\sqrt{\frac{m^2}{{L_1}^2}+\frac{n^2}{{L_2}^2}}},$$

and recall that the periods for a string of length L are $\frac{2L}{nc}$. In both cases, this means that if we know the periods of the harmonics (and speed of propagation c), we can work out the length of the string or the dimensions of the rectangular membrane. But is this generally the case? If we know the periods of a membrane's harmonics, can we deduce the shape of the membrane? This is a difficult question which wasn't answered until 1992 and then in the negative.

Note that a rectangular drum with length 1 and width 2 has the same periods as one with length 2 and width 1, but two such drums that are rotations of one another are considered to have the same shape.

However there are homophonic drums with *different* shapes (Figure 52). 'Homophonic' here means that their harmonics have the same periods. Note that the two drums have the same area; this is not coincidental, as the area of a drum *can* be deduced from the harmonics' periods.

The above is an example of an *inverse problem* in mathematics: given the observable behaviour of a system, can we infer the underlying mechanisms that cause that behaviour? This topic is a massively important area of science generally where a study of the inner working of a system is practically impossible—for example, learning about the Earth's core from studies of how seismic waves travel through the Earth.

The spectral theorem

The harmonics for a vibrating string have a relatively simple form, as do those for the rectangular membrane, being expressible using trigonometric functions. But it is a lot less clear what we might mean by the harmonics of a circular drum or those of even more exotic shapes as in Figure 52.

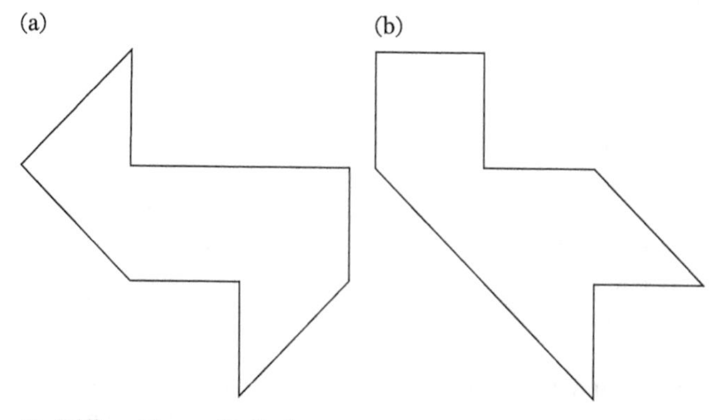

52. Different homophonic drums.

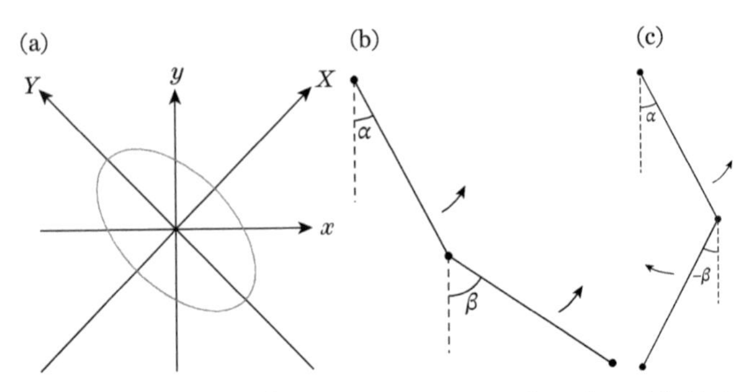

53(a). $\mathbf{4x^2 + 4xy + 4y^2 = 1}$**. 53(b). Slow normal mode. 53(c). Fast normal mode.**

Digressing briefly, consider what the curve with equation $4x^2 + 4xy + 4y^2 = 1$ looks like. One way of approaching this is to rewrite the equation as

$$3(x+y)^2 + (y-x)^2 = 1,$$

or as $3X^2 + Y^2 = 1$, where $X = x + y$ and $Y = y - x$. In these new co-ordinates, X and Y, the curve has the equation of an ellipse (as sketched in Figure 53(a)); sketched also are the new X- and Y-axes. Note that in the new co-ordinates the equation has no 'mixed' XY term and that the new X- and Y-axes are still perpendicular.

The expression $4x^2 + 4xy + 4y^2$ is called a *quadratic form* in two variables, x and y, and such forms can naturally be generalized to three or more variables. In 1829 Cauchy proved the **spectral theorem**, which shows that, for any quadratic form, new variables $X, Y, Z, \ldots$, can be introduced to eliminate all the mixed terms—so only terms like $X^2, Y^2, Z^2, \ldots$, remain—and, further, the new X-,Y-, Z-axes etc. will still be perpendicular.

The spectral theorem is important and widely applicable. For example, for small oscillations of a double pendulum (Figure 53) about its equilibrium (where both pendula hang vertically), its gravitational potential energy is a quadratic form in the angles α and β. By the spectral theorem, new co-ordinates can be found and special *normal modes* of oscillation (akin to the previous harmonics) can be identified. Every small oscillation would be expressible as a sum of these two normal modes. There is a slow mode when the rods move together (Figure 53(b)) and a fast mode when they move in opposite directions (Figure 53(c)).

One problem is that these examples involve finite variables, whilst we have seen that a taut string has *infinitely* many harmonics. What is needed is an infinite-dimensional version of the spectral theorem, and such does exist; that theorem is within the area of **functional analysis**—essentially infinite-dimensional analysis—which developed in the first half of the 20th century through the work of Stefan Banach, David Hilbert, and John von Neumann.

When a string's vibration $y(x,t)$ is expressed as an infinite sum of the harmonics (or normal modes)

$$y_n(x,t) = A_n \sin\left(\frac{n\pi x}{L}\right)\cos\left(\frac{n\pi ct}{L}\right),$$

the energy in the string is an infinite quadratic form involving the square terms A_n^2 but no mixed terms $A_m A_n$, where m and n are distinct. Within this setting, the string's harmonics (which are functions) are still perpendicular to one another in a technical sense; this perpendicularity manifests in the vanishing of integrals such as

$$\int_0^L \sin\left(\frac{m\pi x}{L}\right)\sin\left(\frac{n\pi x}{L}\right)\mathrm{d}x = 0,$$

when m and n are distinct. This is akin to the fact that two vectors $\mathbf{u} = (u_1, u_2, u_3)$ and $\mathbf{v} = (v_1, v_2, v_3)$ in three dimensions are perpendicular if

$$u_1v_1 + u_2v_2 + u_3v_3 = 0.$$

(The expression on the left is the *scalar product* of **u** and **v**.) The previous integral is just a continuous sum equivalent of this. It was through using this 'perpendicularity' that Fourier was able to determine his Fourier coefficients (see Appendix for details).

Distributions

Reality is complicated, so mathematical models make simplifying, but reasonable, assumptions. Such assumptions lead to simpler equations that can be solved, or more easily analysed. Common simplifying assumptions include *point charges* and *point masses* in electromagnetism and gravity. A subatomic particle, with non-zero mass, might be modelled as a point, which has zero volume; given that an electron has mass

$$0.00000000000000000000000000000091 \text{ kg},$$

this might seem a reasonable simplification. However, it would still mean, as a point mass, that this particle has infinite density, because its non-zero mass is contained within zero volume.

Seemingly wilder simplifications get made. For example, a proof that the Earth moves around the Sun in an elliptical orbit would treat both the Sun and Earth as point masses, despite the Sun having mass

$$1,989,000,000,000,000,000,000,000,000,000 \text{ kg}.$$

However, this simplification is still reasonable, as it can be mathematically shown that the gravitational field outside a uniform, spherical body is exactly the same as it would be if all that mass were at the body's centre.

Still, these simplifications can lead to difficulties, particularly if we wish to refer to density. If we consider a point mass of mass 1 at

$x = 0$ on the real number line, then the density of matter $\delta(x)$ satisfies the following:

$$\delta(x) = 0 \text{ when } x \neq 0, \qquad \int_{-\infty}^{\infty} \delta(x)\mathrm{d}x = 1.$$

The first equation states there is no matter except at the origin 0. In the second equation, the integral of the density is the total mass on the whole real line, which we know is 1. *The problem is that no function* $\delta(x)$ *has these properties!* The 'function' $\delta(x)$ is called the **Dirac delta function**, named after Paul Dirac, the theoretical physicist who introduced it in 1930 in his influential book *The Principles of Quantum Mechanics*. But at that time its status as a mathematical object was still unclear.

Nowadays, the Dirac delta function is fully understood as a **Schwartz distribution** or **generalized function**. Laurent Schwartz developed the theory of distributions in the late 1940s, for which he won a Fields Medal in 1950. Distribution theory now sits within the field of functional analysis.

In contrast with functions, which need specifying at every point, distributions might be thought of as having well-defined local averages. This makes them useful in modelling real-world situations. For example, when you hold an object and then let it drop, what can be said of the object's acceleration? There is zero acceleration before the drop, and the acceleration is that of gravity after the drop, but what is the acceleration at the instant of the drop? In practice this doesn't matter, but it would if we wished to describe acceleration with a function rather than a distribution.

The acceleration here is discontinuous at the drop, changing from zero to the value of gravity. But astonishingly, *as a distribution*, the acceleration can be differentiated and its derivative is a multiple of the Dirac delta function. None of this makes *any* sense within the traditional context of calculus—differentiable functions are continuous—but within the context of distributions it all makes rigorous sense. Schwartz didn't just tidy up previously unrigorous loose ends; the theory of distributions proved powerful because of the richness of their calculus and the breadth of their applications.

Quantum theory

Quantum theory developed in the early 20th century to describe certain phenomena of subatomic particles which classical mechanics could not penetrate. Foremost among these phenomena, and what gave quantum theory its name, were experiments which showed that particles could have only certain discrete energies, or 'quanta'.

In quantum theory, the state of a particle is described by a wave function $\psi(x,t)$, which satisfies Schrödinger's equation. For the so-called 'particle in a box' model, Schrödinger's equation takes a form such as the following boundary value problem:

$$i\frac{\partial \psi}{\partial t} = -\frac{h}{4\pi m}\frac{\partial^2 \psi}{\partial x^2}, \qquad \psi(0) = \psi(L) = 0,$$

where L is the length of the box, x denotes position, t denotes time, m is the particle's mass, and h is Planck's constant. Here $i = \sqrt{-1}$ and $\psi(x,t)$ are complex, rather than real, numbers. Quantum theory is naturally set in the language of complex numbers, which we are about to meet in Chapter 7.

As with the wave equation, there are certain states (the normal modes) that are special solutions to Schrödinger's equation, each with a different energy. The wave function is, in general, an infinite sum (or superposition) of these different states, but once energy is measured, the wave function collapses to one of these states. The Fourier coefficient of each state relates to the probability that the wave function collapses to that state.

The rise of quantum theory in the early 20th century introduced many counter-intuitive phenomena, and it was mathematics, and in particular functional analysis and the work of John von Neumann, that helped to give the subject rigorous foundations.

Chapter 7
Putting the *i* in analysis

Complex numbers

Quantum theory—physics at the subatomic level—is more naturally described in the language of *complex numbers* than with real numbers. What follows is a description of how mathematicians became interested, albeit hesitantly, in these so-called 'complex' numbers.

A need for complex numbers arose during the Renaissance as new ways were found to solve polynomial equations such as

$$z^3 - 2z^2 - z + 2 = 0.$$

The *degree* of this equation is 3, this being the highest power of z. Our aim is to find all the solutions z. We can see that $z = 1$ is a solution because

$$1^3 - (2 \times 1^2) - 1 + 2 = 0.$$

You might check that $z = 2$ is also a solution and so is $z = -1$. And that's all of them—three solutions: $z = 1, -1, 2$. Other polynomials, though, seem to have no solutions. For example, the degree 2 equation

$$z^2 = -1$$

has no real numbers as solutions. If you take a positive number, z, then its square z^2 is also positive, and so cannot equal -1; if you take a negative number, then its square is also positive; finally, $0^2 = 0$. So no real number z squares to -1.

The ancient Babylonians knew how to find all, if any, positive solutions of a degree 2 equation. But it wasn't until the 16th century that general equations of degrees 3 and 4 were solved by Italian mathematicians. However, a problem they had was that their methods necessitated calculations involving the square roots of negative numbers, even when all the equation's solutions were real numbers. At this time most mathematicians were ill at ease with negative numbers, let alone their square roots. But if they imagined square roots of negative numbers to exist, and if all these disturbing 'imaginary' numbers cancelled out at the end of the calculation, then their methods did indeed yield correct solutions.

If we imagine there to be a solution to the equation $z^2 = -1$, denoted by i, and if it otherwise adds and multiplies like other numbers, then we can solve polynomial equations that we couldn't previously solve. For example, the number $z = 3 + 2i$ satisfies

$$z^2 - 6z + 13 = 0,$$

because

$$\begin{aligned}&(3+2i)^2 - 6 \times (3+2i) + 13\\ &= (9 + 12i + 4i^2) - (18 + 12i) + 13\\ &= 9 + 12i - 4 - 18 - 12i + 13 = 0,\end{aligned}$$

the last line following because $i^2 = -1$. Numbers of the form $x + yi$, where x and y are real numbers and $i^2 = -1$, are called **complex numbers**.

It's a perfectly reasonable question to ask what a number like $3 + 2i$ could possibly signify. A Renaissance mathematician might have asked the same of a negative number—after all, you can't have -30 apples. But we now habitually see negative numbers describing temperatures, deficit bank balances, or x-co-ordinates of points to the left of the y-axis. In a similar manner, particularly during the 19th century, mathematicians began finding complex numbers so useful as to put aside any philosophical qualms. In particular, the **fundamental theorem of algebra** showed that *all* the solutions of *any* polynomial equation are complex numbers; more precisely, a polynomial of degree n has n solutions among the complex numbers (counting any repetitions).

Cauchy

The worth of complex numbers would soon become even more apparent because of complex analysis as developed by the French mathematician Augustin-Louis Cauchy (Figure 54). Cauchy was a titan of 19th-century mathematics who prolifically made contributions across a wide range of topics, and his name is associated with a diverse range of concepts and theorems, arguably more so than any other mathematician. His work ranges from algebra—results relating to symmetry and a first version of the spectral theorem—through to the study of polyhedra, differential equations, and solid mechanics and elasticity.

His seminal *Cours d'Analyse* of 1821 on real analysis included early use of ε-δ arguments—Bolzano has priority in this, though his work went unnoticed—but Cauchy is most remembered for his almost single-handed development of complex analysis during the 1820s.

Complex analysis is one of the most harmonious, yet applicable, of mathematical subjects. That the complex numbers should prove so powerful may seem surprising at this point. Cauchy took the ideas of differentiation and integration for real functions and generalized them, in fairly natural ways, to functions with complex numbers for inputs and outputs. It turns out that the analysis of

54. Cauchy.

such functions is far richer and more powerful than is the case with real functions.

This may only seem interesting in the abstract. However, a problem involving real numbers, which ostensibly has nothing to do with complex numbers, might be generalized and extended to a problem involving complex numbers. At that point a wealth of theorems becomes applicable. This is what the French mathematician Jacques Hadamard meant by: 'The shortest path between two truths in the real domain passes through the complex domain.' This is a powerful, general approach in mathematics: take a problem out of its natural or initial setting and pose it as a related problem in a setting where more theory can be applied. The actual answer we are looking for is typically still a real number, but we may find the desired answer more easily as the real or imaginary part of a complex number.

The complex plane

The real numbers are commonly represented on an infinite real number line with numbers increasing as we move from left to right. However, complex numbers, which have the form $z = x + yi$, are two-dimensional in nature: x is called the *real part* of z and y is called the *imaginary part* of z. So it makes more sense to represent complex numbers using the xy-plane and having, say, the point with co-ordinates $(1, 2)$ represent the complex number $1 + 2i$ (Figure 55(a)).

When represented in this way, the plane is referred to as the **complex plane**. (The complex plane is also commonly referred to as the *Argand diagram*, after Jean-Robert Argand, who in 1806

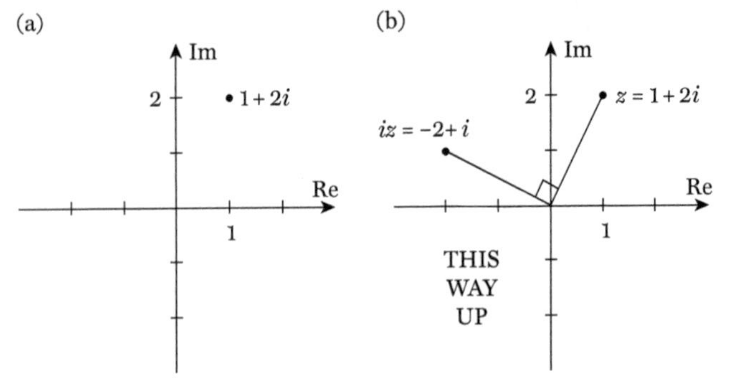

55(a). The complex plane. 55(b). Multiplying by *i*.

represented complex numbers so, though Caspar Wessel did this earlier, in 1799.) Note that the complex plane includes the real line; the x-axis contains numbers of the form $x + 0i = x$, which are just the real numbers, so it is referred to as the *real axis*, and the y-axis is referred to as the *imaginary axis*.

It may seem that the complex plane is just another version of the real xy-plane, but it has an important extra feature that is key to how different the analysis, geometry, and algebra of the complex plane are from those of the real plane. In the complex plane we can multiply by i (and there is no equivalent natural feature in the real plane). If $z = x + yi$, then $iz = -y + xi$. Note that the point iz is at the same distance from the origin as z and is at a right angle anticlockwise around from z (Figure 55(b)). This means that the complex plane comes with an *orientation*, a natural sense of anticlockwise, which the real plane does not. Put another way, the real plane is essentially an infinite sheet of paper with axes drawn on it; by comparison, the complex plane is an infinite sheet of paper with axes *and* a label saying 'THIS WAY UP'. This orientation has important consequences for which transformations of the complex plane are of natural interest in complex analysis.

Two maps of the complex plane

Consider the complex function $p(z) = z^2$. For $z = x + yi$,

$$p(z) = (x + yi)^2 = x^2 + 2ixy + i^2y^2 = (x^2 - y^2) + 2xyi.$$

If we write $u(x, y)$ and $v(x, y)$ for the real and imaginary parts of $p(x + yi)$, then

$$u(x, y) = x^2 - y^2, \quad v(x, y) = 2xy.$$

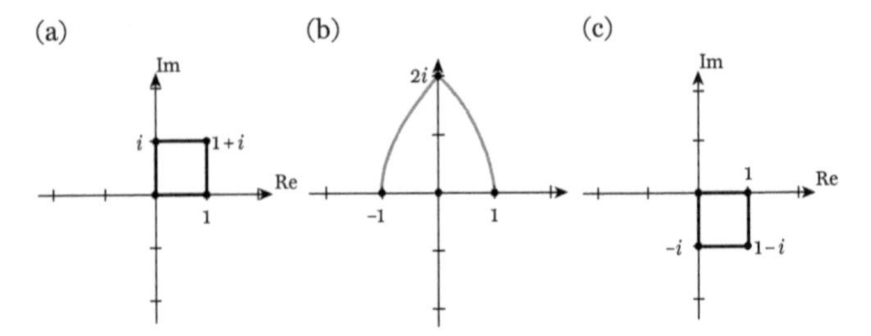

56(a). The unit square. 56(b). Its image under $p(z) = z^2$. 56(c). Its image under $q(z) = \bar{z}$.

Our second map is called *complex conjugation*. Given a complex number $z = x + yi$, its **conjugate** is $q(z) = \bar{z} = x - yi$. Note that $\bar{z}$ is the reflection of z in the real axis, so $q(z)$ flips the complex plane and its 'THIS WAY UP' label. In terms of real and imaginary parts, we have

$$u(x, y) = x, \quad v(x, y) = -y.$$

It is possible to see the effects that $p(z)$ and $q(z)$ have on the unit square (Figure 56). In many ways both these maps are 'nice'; certainly, the partial derivatives $\partial u/\partial x, \partial u/\partial y, \partial v/\partial x, \partial v/\partial y$ all exist. In fact $q(z)$ is particularly nice, as it also preserves distance. However, as we shall soon see, it is precisely $q(z)$ which is problematic when viewed as a map of the complex plane.

Complex differentiability and the Cauchy–Riemann equations

Given a function $f(z)$, which has inputs and outputs which are complex numbers, we say that $f(z)$ is **complex differentiable** if the limit of

$$\frac{f(z+h) - f(z)}{h}$$

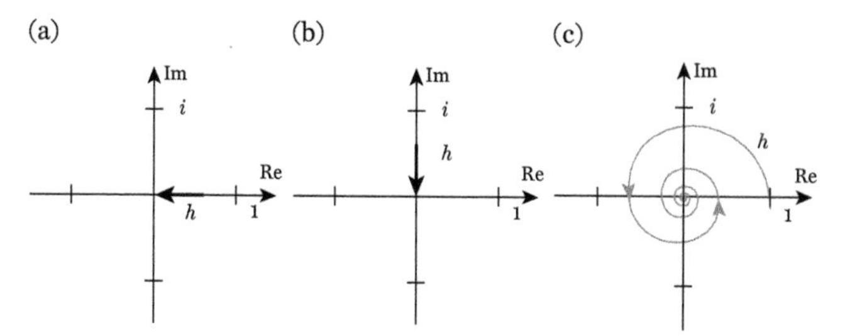

57. Different ways in which the complex number h can become small.

exists as the complex number h becomes small—that is, h converges to 0 in the complex plane. If the above limit exists, then it is again denoted by $f'(z)$ as in Chapter 2, but this limit is a complex number, rather than a real one. So we can't think of the above fraction as the gradient of a chord as we previously did, but it is a natural generalization to complex functions of the limit defining the derivative of real functions from Chapter 2.

As the complex plane is two-dimensional, h can converge to 0 in a variety of ways. Previously, on the real line, a real number h could approach zero essentially 'from the left' (as a small negative number) or 'from the right' (as a small positive number). In the complex plane, these would correspond to h approaching zero along the real axis (Figure 57(a)), and in this case the above limit would equal $\partial f/\partial x$, where x is the real part of z. But there are many other ways in which h can become small (Figures 57(b), 57(c)). The important point here is that we are requiring the same limit $f'(z)$ to exist, *however* h becomes small; we will see this is more demanding than just insisting that the partial derivatives $\partial f/\partial x$ and $\partial f/\partial y$ both exist.

Consider now whether the functions $p(z) = z^2$ and $q(z) = \bar{z}$ are complex differentiable. Note

$$\frac{p(z+h)-p(z)}{h} = \frac{(z+h)^2 - z^2}{h} = 2z + h,$$

which has limit $2z$ as h becomes small in any way whatsoever. The algebra involved is identical to that seen in Chapter 2 when we differentiated x^2.

However, $q(z)$ is *not* complex differentiable. Note that

$$\frac{q(z+h)-q(z)}{h}=\frac{\overline{z+h}-\bar{z}}{h}=\frac{\bar{z}+\bar{h}-\bar{z}}{h}=\frac{\bar{h}}{h}.$$

The conjugate of a complex number is its reflection in the real axis. So if h is real, then it equals its conjugate, giving $\bar{h}/h=1$, and we get a limit of 1 as h approaches 0 along the real axis. But if h is a purely imaginary number (that is, on the imaginary axis), then $\bar{h}=-h$, and so $\bar{h}/h=-1$, giving a limit of -1. Because these are different answers, this means that the $\bar{h}/h$ doesn't have a limit as h becomes small.

So conjugation is not a 'nice' (= complex differentiable) function; this may seem surprising, as conjugation is just reflection in the real axis. The problem is that conjugation does not respect the 'THIS WAY UP' label on the complex plane and turns the plane over, reversing its orientation.

As commented earlier, being complex differentiable is more restrictive than in the real case. If $u(x,y)$ and $v(x,y)$ are the real and imaginary parts of a complex differentiable function $f(x+yi)$, then it can be shown that

$$f'(z)=\frac{\partial u}{\partial x}+\frac{\partial v}{\partial x}i,$$

if h becomes small along the real axis. But if h becomes small along the imaginary axis, we get

$$f'(z)=\frac{\partial v}{\partial y}-\frac{\partial u}{\partial y}i$$

(see the Appendix for details). Because $f(z)$ is complex differentiable, these two limits *must* be the same, and hence these two expressions have equal real parts and equal imaginary parts. This gives us the **Cauchy–Riemann equations**,

$$\frac{\partial u}{\partial x} = \frac{\partial v}{\partial y}, \qquad \frac{\partial v}{\partial x} = -\frac{\partial u}{\partial y}.$$

We can check the Cauchy–Riemann equations hold for $p(z)$, where we have

$$u(x,y) = x^2 - y^2 \qquad \text{and} \qquad v(x,y) = 2xy,$$

so that

$$\frac{\partial u}{\partial x} = 2x = \frac{\partial v}{\partial y} \qquad \text{and} \qquad \frac{\partial v}{\partial x} = 2y = -\frac{\partial u}{\partial y}.$$

However, for $q(z)$ we have $u(x,y) = x$ and $v(x,y) = -y$, and we see

$$\frac{\partial u}{\partial x} = 1 \neq -1 = \frac{\partial v}{\partial y}.$$

This again shows that conjugation is not complex differentiable. In practice, a function satisfying the Cauchy–Riemann equations is complex differentiable. I write 'in practice', as some further mild technical constraints also need to be satisfied. But, broadly speaking, we can think of a complex differentiable function $f(x+yi) = u(x,y) + v(x,y)i$ as one with continuous partial derivatives $\partial u/\partial x, \partial u/\partial y, \partial v/\partial x, \partial v/\partial y$ which satisfy the Cauchy–Riemann equations. By comparison, a function from the real plane to the real plane is differentiable if these four partial derivatives exist and are continuous; this is much less of a requirement, and one which *is* satisfied by $q(x+yi) = x - yi$.

Holomorphic functions

Complex differentiable functions defined on a region of the complex plane are called **holomorphic** or **analytic** functions, but, rather than introduce further technical language, we shall continue referring to them as 'complex differentiable'.

The theory of complex analysis is far richer than that of real analysis. Because we have been more restrictive in our definitions, there is a payoff in that we can prove a lot more. However, complex differentiable functions are still plentiful: they include all functions that we are likely to be interested in, such as polynomials, trigonometric functions, and exponentials, and, with a little care in their definitions, logarithms, roots, and powers are also complex differentiable (see Appendix). So the strength of complex analysis comes from being able to apply a richer theory to a still broad set of functions.

For a function $f(z)$ to be complex differentiable, its derivative $f'(z)$ must exist. For real functions the story might stop there—a real function $f(x)$ might be once but not twice differentiable. Such a function is

$$f(x) = \begin{cases} x^2 & \text{if } x \geqslant 0; \\ -x^2 & \text{if } x < 0. \end{cases}$$

It has derivative $f'(x) = 2|x|$, and we saw in Chapter 2 that the modulus function is not differentiable. However, the derivative of a complex differentiable function is always itself complex differentiable; this means that a complex differentiable function can be differentiated repeatedly.

In fact, complex differentiable functions are even nicer than this: they are also *analytic*, meaning they can be expressed, locally at least, by their Taylor series; we noted in Chapter 3 that this is not generally true of real functions, even those that can be repeatedly differentiated.

Complex trigonometric functions and the exponential function

In Chapter 3 we saw how the sine, cosine, and exponential functions can be defined for real inputs by power series. These same power series also converge when the input is a complex number. In important ways, these complex versions of sine, cosine, and the exponential function are very different from their real counterparts, but using complex numbers, we can appreciate a deep connection between these three functions in the form of *Euler's identity*.

We saw earlier that the power series for sine, cosine, and the exponential function are

$$\sin(z) = z - \frac{z^3}{3!} + \frac{z^5}{5!} - \frac{z^7}{7!} + \frac{z^9}{9!} - \cdots,$$
$$\cos(z) = 1 - \frac{z^2}{2!} + \frac{z^4}{4!} - \frac{z^6}{6!} + \frac{z^8}{8!} - \cdots,$$
$$e^z = 1 + z + \frac{z^2}{2!} + \frac{z^3}{3!} + \frac{z^4}{4!} + \frac{z^5}{5!} + \cdots.$$

These three series converge for all complex numbers z and define complex differentiable functions on the entire complex plane. Again, the derivative of $\sin(z)$ is $\cos(z)$ and the derivative of e^z is e^z. Other identities such as

$$(\cos(z))^2 + (\sin(z))^2 = 1$$

still hold true for complex z. However, this no longer implies that $\sin(z)$ and $\cos(z)$ lie between -1 and 1, as the squares of complex numbers need not be positive. In fact, the complex functions $\sin(z)$ and $\cos(z)$ are unbounded functions that achieve all possible complex numbers as outputs.

On the real line the two trigonometric functions appear unrelated to the exponential. The latter is unbounded and not periodic, unlike the former, but they are intimately related when seen through a complex lens. Firstly, note that the powers of i proceed as follows:

$$i^1 = i, \quad i^2 = -1, \quad i^3 = -i, \quad i^4 = 1, \quad i^5 = i, \quad i^6 = -1,$$

repeating as $i, -1, -i, 1, i, -1, -i, 1, \ldots$ with period four. So,

$$\begin{aligned} e^{iz} &= 1 + iz + \frac{i^2 z^2}{2!} + \frac{i^3 z^3}{3!} + \frac{i^4 z^4}{4!} + \frac{i^5 z^5}{5!} + \frac{i^6 z^6}{6!} + \cdots \\ &= 1 + iz - \frac{z^2}{2!} - i\frac{z^3}{3!} + \frac{z^4}{4!} + i\frac{z^5}{5!} - \frac{z^6}{6!} - \cdots . \end{aligned}$$

If we separate out the terms with or without an i, we obtain

$$e^{iz} = \left(1 - \frac{z^2}{2!} + \frac{z^4}{4!} - \cdots\right) + i\left(z - \frac{z^3}{3!} + \frac{z^5}{5!} - \cdots\right),$$

or, more succinctly,

$$e^{iz} = \cos(z) + i\sin(z).$$

This is **Euler's identity**. In particular, if we set z to be π, and remember we are using radians, we get

$$e^{i\pi} = \cos(\pi) + i\,\sin(\pi) = -1 + 0i = -1.$$

This, particularly when written as $e^{i\pi} + 1 = 0$, is often rated as one of the most beautiful equations in mathematics, as it connects the fundamental numbers $0, 1, \pi, e, i$.

Taylor series and Laurent series

Consider now the function

$$g(z) = \frac{1}{1 - z},$$

where z is a complex number. $g(z)$ is differentiable for all complex numbers, except when $z = 1$, where the function is not even defined (because we cannot divide by 0). This point, 1, where $g(z)$ is not differentiable, is called a **singularity**. We will shortly see that these singularities play a key role in Cauchy's theory of complex integration.

Complex differentiable functions are analytic. So around any point where $g(z)$ is differentiable, $g(z)$ equals its Taylor series. In Chapter 2 we determined the Taylor series for $g(z)$ about 0, namely

$$g(z) = 1 + z + z^2 + z^3 + \cdots.$$

This series doesn't converge for all values of z, though; it converges for those z inside the circle with centre 0 and radius 1 (Figure 58(a)) and does not converge outside that circle. And it is no coincidence that the singularity $z = 1$ is on this circle. Complex differentiable functions equal their Taylor series up to the nearest singularity.

$g(z)$ is also complex differentiable at i, and so $g(z)$ equals its Taylor series centred at i, which converges within the circle centred at i

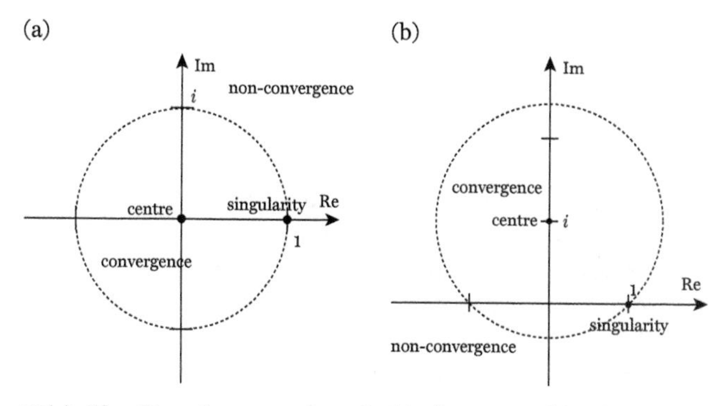

58(a). The disc of convergence of $g(z)$ about 0. 58(b). The disc of convergence of $g(z)$ about i.

and passing through the nearest singularity 1 (Figure 58(b)). Specifically, that Taylor series is

$$g(z) = \frac{1}{1-i} + \frac{(z-i)}{(1-i)^2} + \frac{(z-i)^2}{(1-i)^3} + \frac{(z-i)^3}{(1-i)^4} + \cdots.$$

(Again the first coefficient is $g(i)$, the second is $g'(i)$, etc., as with real Taylor series.) And if we return to the series which define the exponential, sine, and cosine functions, then these are their Taylor series centred at 0. Because these three functions are differentiable at *every* point, they have no singularities. As there is no 'nearest singularity', these series converge on the entire complex plane.

A complex differentiable function has a Taylor series centred at any point where it is differentiable. But what can be said of a function centred at a singularity, where the function isn't differentiable? Certainly the function cannot be expressed as a Taylor series, because power series define differentiable functions, but we can resolve this if we're willing to include negative powers in the series. Such a series is called a **Laurent series**, after the French mathematician Pierre Alphonse Laurent.

As examples, consider

$$\frac{e^z}{z^2} = \frac{1}{z^2} + \frac{1}{z} + \frac{1}{2!} + \frac{z}{3!} + \frac{z^2}{4!} + \cdots,$$

$$e^{1/z} = 1 + \frac{1}{z} + \frac{1}{2!z^2} + \frac{1}{3!z^3} + \frac{1}{4!z^4} + \cdots.$$

The first Laurent series is found by dividing the exponential series for e^z by z^2 and the second is found by substituting $1/z$ into the exponential series. Both functions are differentiable everywhere except when $z = 0$, where each has a singularity.

More generally, Laurent's theorem shows that a function $f(z)$ which has a singularity at $z = a$, but which is otherwise differentiable in the vicinity of a, can be uniquely written as

$$\cdots + \frac{c_{-2}}{(z-a)^2} + \frac{c_{-1}}{z-a} + c_0 + c_1(z-a) + c_2(z-a)^2 + \cdots,$$

where the Laurent coefficients $\ldots, c_{-2}, c_{-1}, c_0, c_1, c_2, \ldots$ are complex numbers. As well as there being potentially infinitely many positive powers, it is entirely possible that there are also infinitely many negative powers. The Laurent series for $g(z)$, centred at its only singularity 1, is fairly straightforward to find, because

$$g(z) = \frac{1}{1-z} = \frac{-1}{z-1}.$$

By the uniqueness of a Laurent series, this means that $c_{-1} = -1$, and as no other power of $z - 1$ is present, all the other Laurent coefficients are zero. We shall see in the next section that the Laurent coefficient c_{-1} is critically important. This coefficient is called the **residue**, for reasons that will also become apparent.

Complex integrals

Integrals in the complex plane are defined in much the same way as we defined line integrals in the real plane (Chapter 5). Let C be a curve in the complex plane which begins at a and ends at b (remember the direction of C is important (Figure 59)), and let $f(z)$ be a complex differentiable function which has real part $u(x, y)$ and imaginary part $v(x, y)$. Then we define

$$\begin{aligned}\int_C f(z)\,\mathrm{d}z &= \int_C (u+iv)(\mathrm{d}x+i\mathrm{d}y) \\ &= \left(\int_C (u\ \mathrm{d}x - v\ \mathrm{d}y)\right) + i\left(\int_C (v\ \mathrm{d}x + u\ \mathrm{d}y)\right).\end{aligned}$$

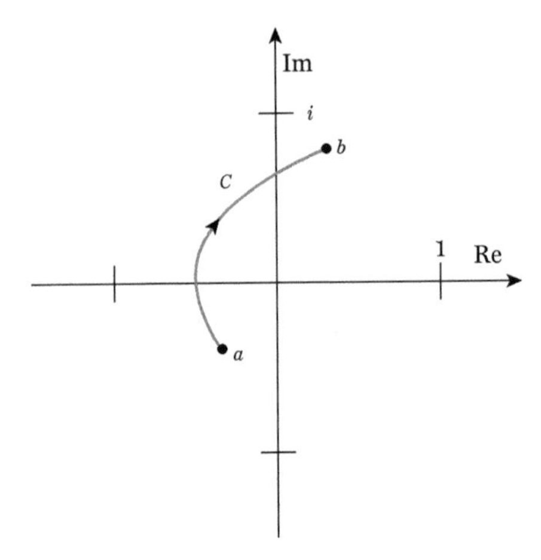

59. A path in the complex plane and its endpoints.

This may look somewhat complicated, but the two integrals in the right expression are the same as those line integrals we discussed in Chapter 5, and together are the real and imaginary parts of a complex number.

Further, familiar results like the fundamental theorem of calculus still apply. If $F(z)$ is a complex differentiable function which has derivative $f(z)$, then

$$\int_C f(z)\, \mathrm{d}z = \int_a^b F'(z)\, \mathrm{d}z = F(b) - F(a).$$

And there is a complex equivalent of Green's theorem known as **Cauchy's theorem** (or *Cauchy's integral theorem*). This states that if C is a loop in the complex plane—a path beginning and ending in the same point—and if $f(z)$ is a function which is complex differentiable *inside and on* the curve C, then

$$\int_C f(z)\, \mathrm{d}z = 0.$$

Ignoring some relatively minor technical details, this theorem is essentially equivalent to Green's theorem (see the Appendix for details).

We can now use these two theorems to evaluate an important complex integral. Our curve C is the circle with centre 0 and radius 1, drawn anticlockwise, and we will evaluate

$$\int_C z^n \, dz,$$

where n is a whole number. If $n \geqslant 0$, then we can use Cauchy's theorem; the function z^n is differentiable inside and on C, and so the above integral equals 0. If $n < 0$, then z^n has a singularity at 0 which is inside C, and so we can't use Cauchy's theorem.

However, if $n \leqslant -2$, we know that z^n has antiderivative $\frac{z^{n+1}}{n+1}$. Considering C as a curve that begins at $a = 1$, goes anticlockwise about 0, and finishes again at $b = 1$, the fundamental theorem gives

$$\int_C z^n \, dz = \frac{b^{n+1}}{n+1} - \frac{a^{n+1}}{n+1} = \frac{1^{n+1}}{n+1} - \frac{1^{n+1}}{n+1} = 0.$$

The only remaining case, and the *only* case of interest, is when $n = -1$ and this integral equals $2\pi i$.

Note (Figure 60) that every point on the circle C can be written as

$$z = \cos(\theta) + i \sin(\theta) = e^{i\theta}.$$

If we consider C as beginning at 1, with $\theta = 0$, then as we go around the circle anticlockwise θ increases until, at $\theta = 2\pi$, we have gone once around C and are back at 1. Note

$$\frac{dz}{d\theta} = -\sin(\theta) + i \cos(\theta) = i(\cos(\theta) + i \sin(\theta)) = ie^{i\theta}.$$

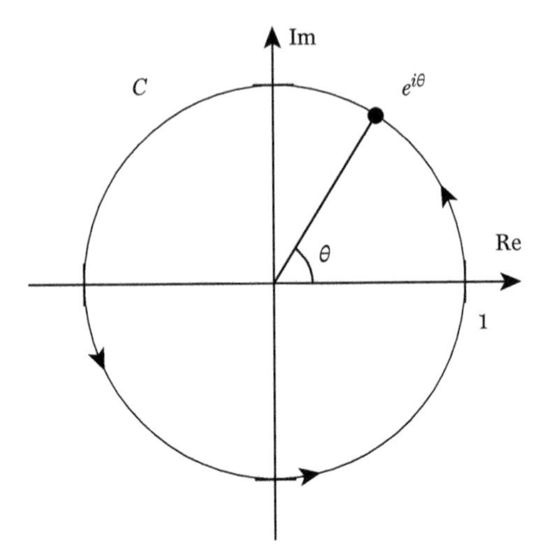

60. Parameterizing the circle.

So we have $z = e^{i\theta}$ and $dz = ie^{i\theta}d\theta$ yielding

$$\int_C z^{-1}\,dz = \int_0^{2\pi} (e^{i\theta})^{-1}(ie^{i\theta})\,d\theta = \int_0^{2\pi} i\,d\theta = 2\pi i.$$

It's not clear yet, but this particular integral is at the heart of many of the theorems of complex analysis. An alternative method of evaluation using the complex logarithm appears in the Appendix.

Cauchy's residue theorem

We are now close to appreciating *Cauchy's residue theorem*, one of the most important theorems of complex analysis and a theorem unlike anything found in real analysis. With C again denoting the anticlockwise unit circle, from the previous calculation we might expect

$$\int_C \left(\cdots + \frac{c_{-2}}{z^2} + \frac{c_{-1}}{z} + c_0 + c_1 z + c_2 z^2 + \cdots \right) \mathrm{d}z = 2\pi i c_{-1}.$$

If we can integrate 'term-by-term', then only the exceptional power z^{-1} leaves any contribution. This is why the coefficient c_{-1} is called the *residue*, being the only coefficient left over.

Justifying 'term-by-term' integration involves some significant analysis; note again that this is a theorem of analysis that shows the order of two limit processes—integrating and summing infinite terms—can be interchanged. But once demonstrated, we see that if $f(z)$ is a function which is complex differentiable inside and on the anticlockwise unit circle C, *except* at 0, then

$$\int_C f(z)\, \mathrm{d}z = 2\pi i \times (\text{residue of } f(z) \text{ at } 0).$$

With a little more work we obtain the following: if C is *any* anticlockwise loop and $f(z)$ is a complex differentiable function inside and on C, except at finitely many singularities, then

$$\int_C f(z)\, \mathrm{d}z = 2\pi i \times (\textit{sum} \text{ of the residues of } f(z) \text{ at the singularities inside } C).$$

This is **Cauchy's residue theorem** at its most general. Note how 'elastic' these integrals are—we could stretch the curve C in various ways and, as long as we don't include new singularities or exclude current ones, the integral won't change.

In a university mathematics course, students learn a wide range of techniques to evaluate real integrals and infinite sums by making appropriate choices for the function $f(z)$ and the curve C. Three such examples that can readily be approached using the residue theorem are

$$\int_0^{2\pi} \frac{dx}{5-4\cos x} = \frac{2\pi}{3}, \quad \int_0^{\infty} \frac{\sin x}{x}\, dx = \frac{\pi}{2}, \quad \frac{1}{1^4}+\frac{1}{2^4}+\frac{1}{3^4}+\frac{1}{4^4}+\cdots = \frac{\pi^4}{90}.$$

These techniques are, in the main, well beyond the scope of this short text. But in general the aim is to reinterpret the above real integrals and infinite sums as complex integrals, or the real or imaginary parts of such integrals, and calculate these using the residue theorem. More detail on how the first integral can be evaluated is in the Appendix.

Conformal maps and applications

We conclude this chapter with another application of complex analysis. Recall that the real and imaginary parts, $u(x,y)$ and $v(x,y)$, of a complex differentiable function $f(z)$ satisfy the Cauchy–Riemann equations. There are two important consequences of this.

One is that the functions $u(x,y)$ and $v(x,y)$ satisfy *Laplace's equation*,

$$\frac{\partial^2 u}{\partial x^2} + \frac{\partial^2 u}{\partial y^2} = 0 = \frac{\partial^2 v}{\partial x^2} + \frac{\partial^2 v}{\partial y^2}$$

(see the Appendix for details), which is a particularly important partial differential equation in the study of fluid dynamics, gravity, and electromagnetism.

The second consequence is that if $f'(z)$ is non-zero, then $f(z)$ is a **conformal** map, meaning that $f(z)$ preserves angles. To some extent we have seen this before, when we looked at the effect of $p(z) = z^2$ on the unit square (Figure 56(b)). The right angles that were at $1, 1+i$, and i remain right angles at $1, 2i$, and -1 in the image; note, though, that the right angle at 0 becomes a half-angle in the image, but this is because $p'(z) = 2z = 0$ when $z = 0$. So $p(z)$ *isn't* conformal at 0.

A more complicated example is the effect that sine has on the semi-infinite strip where $-\pi/2 < x < \pi/2$ and $y > 0$ (Figure 61(a)). The function $\sin(z)$ takes this strip to the upper half of the complex plane (Figure 61(b)). Further, the horizontal and vertical lines are transformed into curves—specifically, ellipses and hyperbolas—but importantly, the right angles in the diagram on the left (Figure 61(a)), between the dashed horizontal and vertical lines, remain as right angles in the right-hand diagram (Figure 61(b)). Note that there are two points on the strip's boundary where angles are not preserved—namely, the right angles at $\pi/2$ and $-\pi/2$ become half-angles at 1 and -1 in the image. These are again points where the derivative of $\sin(z)$, namely $\cos(z)$, is zero and $\sin(z)$ isn't conformal.

The upper half-plane seems a nicer region to work with than the strip; it's possible to envisage how a fluid might flow horizontally in the half-plane (Figure 62(a)). In Figure 62(b) we can see the fluid flow in the strip which $\sin(z)$ transforms to the horizontal flow in the half-plane. More generally, fluid flows in the strip

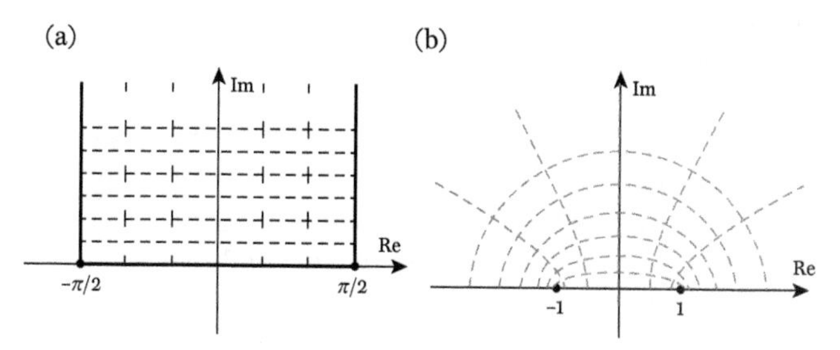

61(a). The semi-infinite strip. 61(b). Its image under $\sin(z)$.

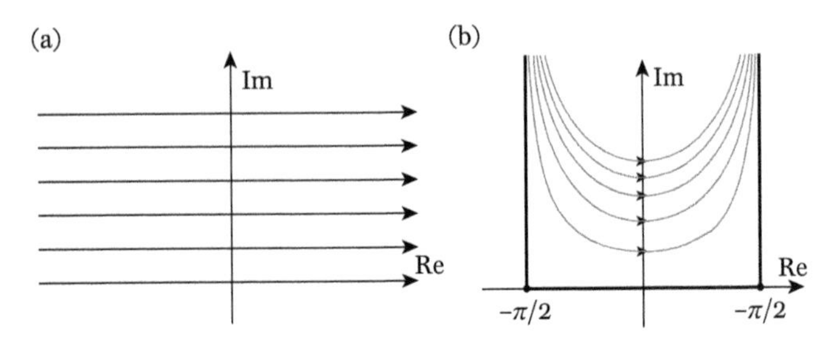

62(a). Fluid flowing horizontally in the half-plane. 62(b). Fluid flowing in the semi-infinite strip.

correspond, using $\sin(z)$, to fluid flows in the half-plane, and vice versa.

We recognize, then, that understanding fluid flows (or solutions to Laplace's equation) in the strip is equivalent to understanding the same problems in the half-plane, a much nicer region. Indeed, this method is particularly powerful in light of the **Riemann mapping theorem**, which states that any region of the complex plane which isn't the whole plane, and which doesn't contain any holes, can be transformed into the half-plane. Solutions to Laplace's equations in the half-plane are well understood, and so, in principle, are solutions in the regions to which the Riemann mapping theorem applies. I write 'in principle', because the maps necessary to transform a region into the half-plane can be computationally messy in practice.

In the early days of flight, Nikolai Joukowski applied such theory to fluid flow around an aerofoil, such as the shape of an aeroplane's wing. Fluid flow around a circular disc is relatively straightforward to model (Figure 63(a)). Joukowski introduced the following complex map:

$$f(z) = z + \frac{1}{z},$$

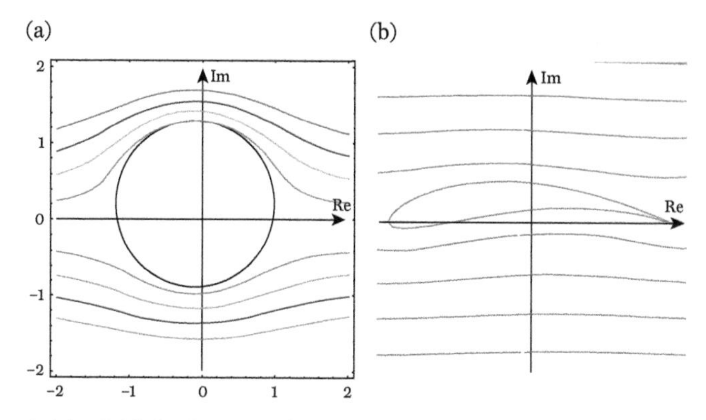

63(a). Fluid flowing around a disc. 63(b). Fluid flowing around an aerofoil.

now named after him, which can transform a disc to the shape of an aerofoil (Figure 63(b)). The Joukowski map then relates the fluid flow about the disc to the fluid flow about the aerofoil, and vice versa. Joukowski was thereby able to explain how lift is generated on an aerofoil.

The study of fluid dynamics and flights has moved on enormously since Joukowski's time, but complex analysis remains a rich and important area of mathematics and science. It has important applications within mathematics—from number theory (more on this in the next chapter) through to profound connections with geometry and topology—and provides an important toolkit in diverse fields such as differential equations, solid and fluid dynamics, acoustics, imaging, and computer vision.

Chapter 8
But there's more . . .

In this final chapter we touch on some further developments in the field of analysis that were not discussed in earlier chapters.

Lebesgue integration

Riemann's integration theory, whilst a significant advance, did not prove adequate for the purposes of 20th-century analysis. It was largely sufficient for calculating the integral of any function that might arise from a real-world problem, but did not prove sufficiently comprehensive and powerful for the needs of mathematics. The difference here is that, in seeking to prove a general theorem in analysis, a mathematician begins with arbitrary integrable functions and may wish to create new functions from them by algebraic and analytic methods. Unfortunately, it is easy to step outside the comfort zone of Riemann integration, producing functions which are not Riemann integrable.

In 1829, Dirichlet introduced a function which is not Riemann integrable. This function is certainly one that wouldn't arise from modelling reality, but is the sort of pathological example which mathematicians need to be concerned with when seeking to prove general theorems. Dirichlet's function $D(x)$ is defined on the interval $0 \leqslant x \leqslant 1$ by

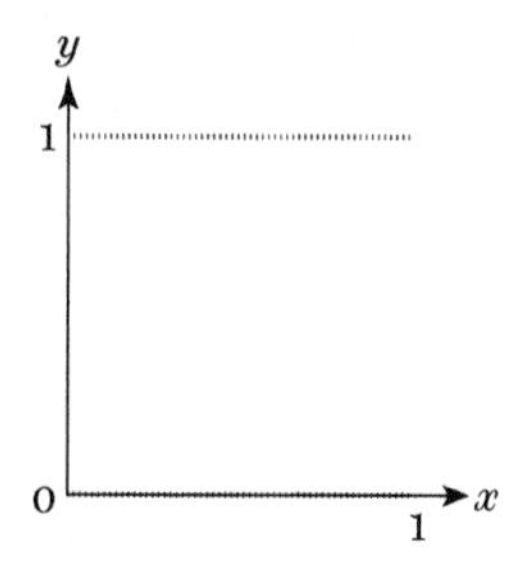

64. Dirichlet's function.

$$D(x) = \begin{cases} 1 & \text{if } x \text{ is rational,} \\ 0 & \text{if } x \text{ is irrational.} \end{cases}$$

Recall that a real number is called rational if it is a fraction of two whole numbers such as $-\frac{1}{2}$ or $\frac{22}{7}$; otherwise, it is termed irrational. The rational numbers are so dense within the real numbers that a graph of $D(x)$ would appear—to the naked eye at least—as two solid lines (Figure 64).

A step function satisfying $\varphi(x) \leqslant D(x)$ can be at most 0, except for finitely many rational x, where $\varphi(x)$ can be as great as 1; this is because every interval of positive length includes irrational numbers. This means that the lower Riemann integral of $D(x)$ equals 0. Likewise, a step function satisfying $\psi(x) \geqslant D(x)$ must be at least 1, except for finitely many irrational x, where $\psi(x)$ can be as small as 0. So the upper Riemann integral equals 1. As these values aren't equal, $D(x)$ is not Riemann integrable.

It may not appear important that such an esoteric function lies outside Riemann's theory; however, whilst $D(x)$ isn't integrable, it *is* the limit of a sequence of integrable functions. The rational numbers are countable and so can be listed as $q_1, q_2, q_3, q_4, \ldots$ A sequence of step functions $\varphi_n(x)$ can then be created by setting $\varphi_n(x) = 1$ at $q_1, q_2, \ldots, q_n$, and otherwise equalling 0. These step functions converge to $D(x)$; we are essentially pushing points from the x-axis up to the graph of $D(x)$ one at a time. The problem is that these step functions are Riemann integrable, but their limit, $D(x)$, *isn't.*

That non-integrable functions can be created so easily from integrable functions is a severe weakness for Riemann's theory. In 1902, Henri Lebesgue (Figure 65(a)) introduced a much more comprehensive theory of integration with powerful convergence theorems. Further, his theory naturally encompassed integrals of unbounded functions and functions on unbounded intervals such as

$$\int_0^1 \frac{\mathrm{d}x}{\sqrt{x}} \quad \text{and} \quad \int_1^\infty \frac{\mathrm{d}x}{x^2},$$

which could only be treated in Riemann's theory as improper integrals—that is, as limits of Riemann integrals.

Lebesgue also proved a result which neatly places Riemann's theory of integration within his own theory: a bounded function $f(x)$ on the interval $a \leqslant x \leqslant b$ is Riemann integrable if and only if its discontinuities form a set of *measure* zero. By contrast, Dirichlet's function is discontinuous on the entire interval $0 \leqslant x \leqslant 1$, which has measure 1.

Measure theory

Lebesgue's theory of integration is grounded in the notion of **measure**, as developed by his tutor Émile Borel, and substantially extended by Lebesgue. Measure is a generalization of length, area, and volume to other dimensions, which encompasses esoteric sets well beyond the familiar. Lebesgue integration comfortably handles Dirichlet's function, because $D(x)$ is 0 except on the rational numbers between 0 and 1, and the rationals form a set of measure zero, also known as a **null set**.

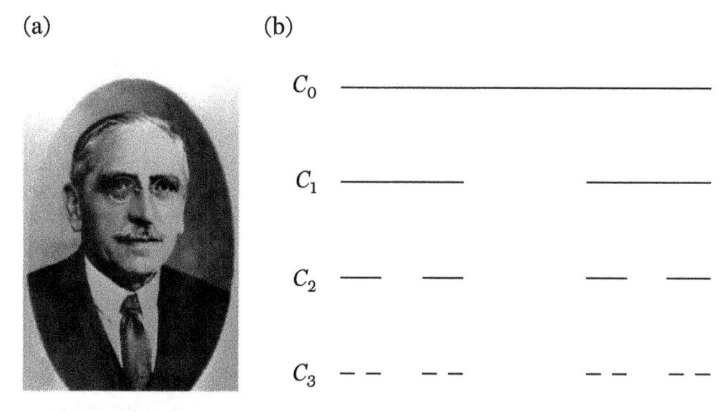

65(a). Lebesgue. 65(b). Constructing Cantor's set.

A subset of the real numbers is a null set if, given any positive number ε, the set can be covered by intervals whose total length is at most ε. This definition captures the fact that the set has 'total length' less than any given positive number, and because measure can't be negative, the only remaining possibility is zero. The null sets are the 'negligible' sets of integration: for example, two functions which differ only on a null set have the same integral. Countable sets, like the rationals, are null (see the Appendix), so, because $D(x)$ differs from the zero function only on a null set, they have the same integral of zero.

There are null sets which are uncountable. One such set is the **Cantor set** C, which is constructed recursively (Figure 65(b)). We begin with C_0 being the whole interval $0 \leqslant x \leqslant 1$. We create C_1 by removing the middle third of C_0, create C_2 by removing the middle thirds of the two intervals making up C_1, and so on. The Cantor set is the collection of points which lie in every C_n. Because C_n comprises 2^n intervals of length $\left(\frac{1}{3}\right)^n$, the measure of C_n equals $\left(\frac{2}{3}\right)^n$. This becomes arbitrarily small as n increases and so C has measure zero.

The study of measure, as a generalization of length, area, volume etc., led to the discovery of some rather pathological examples within mathematics. Given a two-dimensional square with side 1, its area equals 1. If we are asked for the volume of the square, the only reasonable answer seems to be 0; as a three-dimensional object, the square has dimensions $1, 1, 0$ and so volume $1 \times 1 \times 0 = 0$. If asked for its length, one justifiable answer is infinity; the top edge of the square has length 1, but the square itself is made up of infinitely many horizontal line segments, each of length 1.

The square is two-dimensional. What is happening here is that when we evaluate the measure of the square in a higher dimension than 2, we get 0, and in a lower dimension we get infinity. In 1918, Felix Hausdorff generalized the notion of measure to all dimensions, not just to whole numbers. Hausdorff associated with a given set a number $d \geqslant 0$, now known as its *Hausdorff dimension*. The measure of the set in any dimension lower than d is infinite and in any dimension above d it is 0. The measure of the set in dimension d may be finite or infinite. Unsurprisingly, the Hausdorff dimension of a square is 2. More surprisingly, there are sets whose dimension are not whole numbers—such a set is the Cantor set, which has Hausdorff dimension $d = \frac{\log(2)}{\log(3)} = 0.6309\ldots$ The Cantor set's measure is 0 in a higher dimension, infinite in a lower dimension, and 1 in dimension d. The Cantor set is an example of a *fractal*, the word coined to reflect its 'fractional dimension'.

Lebesgue's theory is incredibly comprehensive—Lebesgue measurable sets and functions include any that can be explicitly constructed. In 1905 Giuseppe Vitali showed the existence of a non-measurable set; he used a further axiom of set theory called the *axiom of choice*, which makes statements about the existence of sets in a non-constructive way. Vitali proved that any value assigned for the measure of his set would lead to a contradiction.

Most mathematicians are comfortable assuming the axiom of choice, and view mathematics as too limited without it; this is despite the axiom leading to the existence of non-measurable sets.

Further, in 1933 Andrey Kolmogorov used measure theory to lay the foundations of probability theory, where consequently there are random occurrences to which no probability can be assigned. Analysis would also help provide the technical language for **Brownian motion**, with the development of *stochastic calculus*. Brownian motion is named after the Scottish botanist Robert Brown to describe the random motion of particles in a medium. Random walks, where a particle can move discrete steps at discrete time intervals, are relatively straightforward to model. But there are technical difficulties in describing a continuous version of such random walks by a particle making infinitesimal steps. In 1923 Norbert Wiener gave a mathematical definition of Brownian motion.

A yet more startling result from 1924 is the **Banach–Tarski paradox**, which states that a solid sphere can be broken down into finitely many pieces and these pieces can then be moved and rotated and reassembled as two solid spheres of the same size as the original, thus doubling the volume. This theorem at first seems wholly ridiculous, as we expect that volume should be preserved; however, if the pieces are non-measurable, there is no reason that volume need stay invariant during the decomposition. Perhaps more surprising is that this paradox is true in three dimensions, but not in two. It is possible to assign a measure to all the subsets into which a disc might be finitely decomposed, and so area must remain an invariant of the decomposition. The reason for the difference is algebraic and relates to the interaction of rotations and translations, which is less complicated in two dimensions than it is in three dimensions.

The bar chart (Figure 66) shows what may happen when a fair coin is tossed 20 times. The number of heads achieved is between 0 and 20 (on the horizontal axis) and the bars' heights represent the associated probabilities. Note the bar chart approximates a bell curve, technically a **normal distribution**, also sketched. As more coin tosses are repeated, the bar chart approximates the bell curve

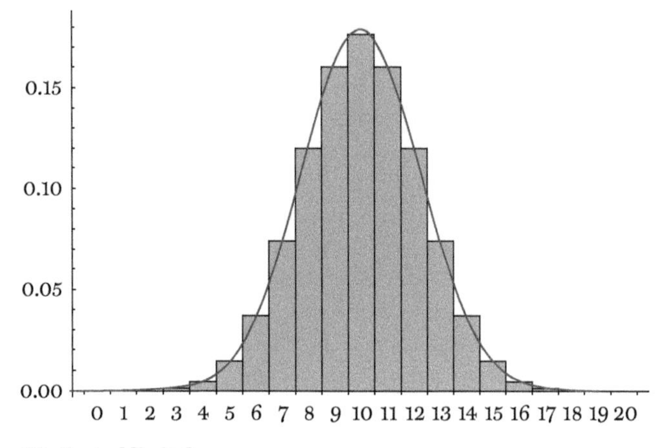

66. Central limit theorem.

better and better. This fact was understood as early as the 18th century by Abraham de Moivre, including in the case where the possibilities (heads or tails) are not equally likely.

Getting one head from a coin toss, or zero if it lands tails, is an example of a *random variable*. More generally, a random variable might be a die roll (a whole number between one and six), the number of people in a bus queue (still a whole number, but now with a larger range), or today's temperature (which has a continuous range of possibilities). If we repeatedly roll a die, then we expect to achieve an average score, long term, of around

$$\frac{1+2+3+4+5+6}{6} = 3.5.$$

We recognize there will be periods of good and bad luck, but in the long term they will cancel out. This fact is precisely captured by the **central limit theorem**, which is one of the most important theorems in probability and statistics. A general version was proved by Aleksandr Lyapunov in 1900–1, which applies generally to all random variables. In the long term, a large sample taken of a random variable will approximate a normal distribution with the expected average and spread. But a precise statement of the theorem, and its proof, are very much analytical in nature.

Analytic number theory

Analytic number theory is usually dated back to **Dirichlet's theorem** of 1837. Recall that a prime number is a whole number $p \geqslant 2$ whose only factors are 1 and p. The list of primes begins

$$2,\ 3,\ 5,\ 7,\ 11,\ 13,\ 17,\ 19,\ 23,\ 29,\ 31,\ 37,\ 41,\ 43,\ 47,\ \ldots,$$

and it has been known from ancient times that there are infinitely many prime numbers (see Appendix). It's a natural generalization of this fact to ask whether there are infinitely many primes among *arithmetic sequences* such as

$$4, 7, 10, 13, 16, 19, \ldots \quad 2, 5, 8, 11, 14, 17, \ldots, \quad 5, 9, 13, 17, 21, 25, \ldots$$

An arithmetic sequence is one that grows with the same common difference each time—this is 3 in the first two examples above and 4 in the third. Clearly, if the first term and the common difference have a factor in common—for example, the first term is 3 and the common difference is 6—then every term after will be divisible by that common factor and not be a prime. But if the first term and common difference have no common factor other than 1, it's plausible that there are infinitely many primes in the sequence.

Euler, in 1737, used some relatively straightforward analysis to prove that the infinite sum of the primes' reciprocals,

$$\frac{1}{2}+\frac{1}{3}+\frac{1}{5}+\frac{1}{7}+\frac{1}{11}+\frac{1}{13}+\frac{1}{17}+\cdots,$$

does not converge. Consequently, there are infinitely many primes; were there only finitely many, then the sum of their reciprocals would be finite. Note that the converse is not true; there may be infinite but suitably sparse numbers whose reciprocals have a finite sum, as with the square numbers or powers of two. Indeed, in 1915, in an effort to prove there are infinitely many *twin primes*—primes two apart such as $17, 19$ and $101, 103$—Viggo Brun considered the sum of their reciprocals only to show this sum is finite. It remains unknown as to whether there are infinitely many twin primes or not.

Dirichlet sought, using analytic methods, to prove the sum of the reciprocals of primes appearing in arithmetic sequences does not converge. For the earlier examples these would be the sums

$$\frac{1}{7}+\frac{1}{13}+\frac{1}{19}+\frac{1}{31}+\cdots, \quad \frac{1}{2}+\frac{1}{5}+\frac{1}{11}+\frac{1}{17}+\cdots, \quad \frac{1}{5}+\frac{1}{13}+\frac{1}{17}+\frac{1}{29}+\cdots.$$

To do this, he introduced what are now known as *Dirichlet series*, infinite sums of the form

$$a_1+\frac{a_2}{2^z}+\frac{a_3}{3^z}+\frac{a_4}{4^z}+\frac{a_5}{5^z}+\cdots,$$

where z is a complex variable and $a_1, a_2, a_3, \ldots$ is a sequence of complex numbers. Dirichlet series remain an important tool. The most important unsolved problem in mathematics is the **Riemann hypothesis**, which concerns the zeros of the Riemann zeta function, which is the Dirichlet series when each a_n equals 1. The hypothesis has important implications for the distribution of the prime numbers.

More recently, in 2004, Ben Green and Terence Tao proved a result of a converse nature to Dirichlet's theorem; they showed the existence of arbitrarily long arithmetic sequences among the prime numbers. The theorem only demonstrates existence and finding such sequences is difficult; in 2019 an arithmetic sequence of 27 primes was discovered and its first term is over 224 quadrillion!

At first glance, the prime numbers appear to arise randomly. There is a sense they get scarcer, but I doubt whether anything seems immediately apparent. By studying tables of prime numbers, various mathematicians—Euler, Gauss, Legendre—were led to conjecture that, for large n, the number of primes less than n is approximately

$$\frac{n}{\log(n)}.$$

Loosely speaking, this says that the chance of a number n being prime is $1/\log(n)$, which decreases as n increases. This fact is now called the **prime number theorem**—its proof was another success for analytic number theory. In 1896 Jacques Hadamard and Charles-Jean de la Vallée Poussin independently proved the theorem, each using complex analytic methods.

Hyperreal numbers

The hyperreals are an extension of the real numbers that allow a rigorous treatment of infinities and infinitesimals. They were developed in 1966 by Abraham Robinson in his work *Non-Standard Analysis*, which built on earlier work of Edwin Hewitt from 1948.

A hyperreal number can be represented by a sequence of real numbers such as

$$(1, 2, 3, 4, \ldots), \quad \left(1, \frac{1}{2}, \frac{1}{3}, \frac{1}{4}, \ldots\right), \quad (1, 1, 1, 1, \ldots).$$

A real number x is identified with the constant sequence $(x, x, x, x, \ldots)$, so that the third hyperreal above is identified with 1.

The first hyperreal above is an infinity—because the sequence tends to infinity—and the second hyperreal is an infinitesimal—because the sequence converges to 0. Hyperreals add and multiply termwise, so the product of the first two hyperreals equals the third.

Unfortunately, there are considerable technical issues with this approach. Firstly, there are just too many sequences; for example, it quickly becomes apparent that the sequence $(0, 1, 1, 1, \ldots)$ also represents 1. So, two sequences are defined as representing the same hyperreal number if 'most' of their terms agree: this notion of 'most' is again technical and involves introducing a measure on the counting numbers; each subset of the counting numbers is given measure 0 or 1, and a subset contains 'most' of the counting numbers if it has measure 1. Two hyperreals can then be ordered if most of the terms of one hyperreal are less than those of the second hyperreal.

A second issue arises from products like

$$(1, 0, 1, 0, \ldots) \times (0, 1, 0, 1, \ldots) = (0, 0, 0, 0, \ldots).$$

In a field, two non-zero elements cannot multiply to give zero. The first hyperreal agrees with $(1, 1, 1, 1, \ldots)$ for odd terms and with $(0, 0, 0, 0, \ldots)$ for even terms. Depending on our choice of 'most', either the set of odd numbers has measure 1 and the set of even numbers has measure 0, or vice versa. In the first case, this means that the first hyperreal is identified with 1 and the second with 0 and the above product says nothing more startling than $1 \times 0 = 0$.

Within this framework it is then possible to define what an infinitesimal or infinite hyperreal is. This in turn opens an entirely different treatment of calculus. Any finite hyperreal can be uniquely written as $a + \varepsilon$ where a is real, known as its *standard part*, and ε is an infinitesimal. A real function $f(x)$ can be naturally extended to a function on the hyperreals, and f can be shown to be differentiable if

$$\frac{f(x+\varepsilon)-f(x)}{\varepsilon}$$

has the same standard part for all infinitesimals ε, that standard part equalling $f'(x)$. Seemingly, after centuries of looking to avoid infinitesimals, the story of analysis had come full circle.

Epilogue

The history of analysis teaches us that mathematics is its own subject (or perhaps, instead of 'mathematics', that should read 'mathematics as done by humans'). In a mannerly fashion, by and large, mathematics has listened to and been influenced by its friends—physics, philosophy, computer science, etc.—but has ultimately come to its own conclusions. The notion of function developed through its use in modelling real-world situations, but the modern notion of function is much more advanced than everyday situations require and has ultimately been developed to serve the purposes of mathematics. Even when analysis is being used to solve a real-world problem, a proof may require theory to be developed well beyond the scope of the problem itself. Further, there is a pure instinct for mathematicians to generalize widely, something which sets them apart from scientists studying this universe.

In undergraduate mathematics degrees, analysis is the course where students learn their trade of rigorous argument. More than any other course, it makes clear the step up to higher education, as students are made to justify their arguments in detail, or provide counter-examples to claims they previously thought were obviously true. In due course, however, students come to appreciate the rigour that analysis courses instil.

To quote the mathematical historian Morris Kline, 'development must be preceded by a period of free, numerous, disconnected, often accidentally discovered, and perhaps disordered creations'. As we have seen in much of this text, whether grappling with infinity, handling Fourier's new approach to modelling heat, making explicit the notions of point masses, or using Von Neumann's drawing together of different approaches to quantum theory, mathematical analysis has commonly provided some certainty, closure, and confidence—secure foundations upon which future generations of mathematicians could build.

Appendix

Collected in this appendix are a number of asides and some results that were considered too calculation-heavy for the main text. But hopefully they are approachable and of interest to a number of readers.

Chapter 1

Ramanujan's approximation to π

In 1910 Ramanujan gave the following approximation to π. Using the first term gives π to six decimal places, with every subsequent term providing eight more decimal places.

$$\frac{9801/\sqrt{8}}{\left(1103+\frac{4!(1103+1\times 26390)}{(1!)^4 396^4}+\frac{8!(1103+2\times 26390)}{(2!)^4 396^8}+\frac{12!(1103+3\times 26390)}{(3!)^4 396^{12}}+\cdots\right)},$$

where $k! = 1 \times 2 \times 3 \times \cdots \times k$.

Cantor's proof that the real numbers are uncountable

We will prove that the real numbers are uncountable by showing that there are uncountably many real numbers in the interval $0 \leqslant x < 1$. The proof shows that any list of numbers from this range misses some real number and so cannot be complete.

Let $x_1, x_2, x_3, x_4, \ldots$ be an infinite list of such numbers, with their decimal expansions. For example, a list might begin

$$\begin{aligned} x_1 &= 0.\mathbf{2}78190632\ldots \\ x_2 &= 0.2\mathbf{1}5249718\ldots \\ x_3 &= 0.81\mathbf{5}311166\ldots \\ x_4 &= 0.957\mathbf{2}85101\ldots \end{aligned}$$

Our task is to create a number x in the interval $0 \leqslant x < 1$ which is not on this list. Or, put another way, x must not equal x_1, x must not equal x_2, x must not equal x_3, and so on.

We can make sure that x does not equal x_1 by having the decimal expansion of x begin with 0.5 rather than 0.2. A 5 rather than a 2 in the tenths column (highlighted in bold) ensures that x does not equal x_1. We then use the hundredths column to make sure x does not equal x_2. We might continue the decimal expansion of x as 0.53, so that the second decimal digit of 3 is different from that of x_2, which is 1 (again in bold).

And we continue in this fashion, at the nth stage choosing the nth decimal place of x to be different from the (bold) nth decimal place of x_n and thereby ensuring that x does not equal x_n. In this way we see that x is nowhere on the list; we can do this for any list, and so no list can be complete.

Because this interval alone cannot be counted, the real numbers can't all be counted either. Cantor gave this simpler proof in 1890, his first proof dating back to 1874.

The axioms of the real numbers

The real numbers are a set, denoted by $\mathbb{R}$, together with the operations of addition $+$ and multiplication $\times$ and a relation $<$ 'is less than', providing a notion of order, which satisfy the following axioms (= assumed rules).

- **Field Axioms**

1. $+$ is associative: $x+(y+z)=(x+y)+z$ for any x, y, z.
2. $+$ is commutative: $x+y=y+x$ for any x, y.
3. There is an element 0 such that $x+0=x$ for all x.
4. For any real number x there is y such that $x+y=0$.
5. $\times$ is associative: $x \times (y \times z)=(x \times y) \times z$ for any x, y, z.
6. $\times$ is commutative: $x \times y=y \times x$ for any x, y.
7. There is a second element 1 such that $x \times 1=x$ for all x.
8. For any non-zero, real number x there is y such that $x \times y=1$.
9. $\times$ distributes over $+$: $x \times (y+z)=x \times y+x \times z$ for any x, y, z.

Any set, with operations of $+$ and $\times$, satisfying these nine axioms is called a **field**. The real numbers, the rational numbers, the complex numbers (Chapter 7) and hyperreals (Chapter 8) are all examples of fields. The integers don't meet axiom 8 as there is no y when $x=2$.

- **Order axioms**

1. For any x, y, if $0<x$ and $0<y$, then $0<x+y$.
2. For any x, y, if $0<x$ and $0<y$, then $0<x \times y$.
3. For any x, precisely one of the following is true:

$$0<x \quad \text{or} \quad 0=x \quad \text{or} \quad x<0.$$

Any field with an order relation $<$ which satisfies these three axioms is called an **ordered field**. The real numbers, the rational numbers, and the hyperreals are ordered fields. But no such order $<$ exists on the complex numbers.

- **Completeness axiom**

Any bounded, increasing sequence of real numbers $x_1<x_2<x_3<x_4<\cdots$ converges to a limit.

The rational numbers do not satisfy the completeness axiom. For example, consider the increasing sequence

$$3, \quad 3.1, \quad 3.14, \quad 3.141, \quad 3.1415, \ldots$$

of decimal approximations of π. Because the real numbers satisfy the completeness axiom, this sequence must have a limit, namely π. The above is an increasing sequence of rational numbers which lie between 3 and 4, and which, because π is irrational, doesn't have a limit among the rational numbers.

The real numbers can be described by other sets of axioms, but these different approaches are ultimately equivalent. Alternatively, some authors might begin with axioms for the counting numbers, deducing the properties of the rational numbers and the real numbers as theorems.

Chapter 2

The equation of the cissoid of Diocles

Referring to Figure 4(b), a general point R on the tangent line has co-ordinates $(t, 2a)$. The point Q lies on the line of OR, and so has co-ordinates $(ct, 2ca)$, where c is chosen so that Q is at distance a from the circle's centre $(0, a)$. This means that

$$(ct)^2 + (2ca - a)^2 = a^2.$$

Solving this gives $c = \frac{4a^2}{t^2 + 4a^2}$. The point P lies on the line OR such that OP equals QR, so that $P = ((1 - c)t, 2(1 - c)a)$. The co-ordinates of P are then

$$x = \frac{t^3}{t^2 + 4a^2}, \quad y = \frac{2at^2}{t^2 + 4a^2}.$$

This is a parametric description of the cissoid in terms of t. But we can eliminate t by noting $t = \frac{2ax}{y}$; substituting this expression for t into either equation and rearranging gives

$$(x^2 + y^2)y = 2ax^2.$$

The rules of differentiation

For differentiable functions $f(x)$ and $g(x)$ and a constant c, the following rules hold:

- $f(x) + g(x)$ is differentiable with derivative $f'(x) + g'(x)$.
- $cf(x)$ is differentiable with derivative $cf'(x)$.
- $f(x)g(x)$ is differentiable with derivative $f'(x)g(x) + f(x)g'(x)$.
- $f(g(x))$ is differentiable with derivative $f'(g(x))g'(x)$.
- $f(x)/g(x)$ is differentiable with derivative

$$\frac{g(x)f'(x) - f(x)g'(x)}{g(x)^2}.$$

The last three rules are referred to as the product rule, the chain rule, and the quotient rule; the quotient rule requires $g(x)$ to be non-zero for $f(x)/g(x)$ to be defined. These rules are commonly credited to Leibniz who was aware of them by 1677.

Chapter 3

Basic identities of the exponential and logarithmic functions

1. For a fixed but arbitrary real number a, applying the product and chain rules shows the function $f(x) = e^{x+a}e^{-x}$ has zero derivative and so is constant. As $f(0) = e^a$, then $e^{x+a}e^{-x} = e^a$ for all x, a. Replacing a with $x + a$ and x with $-a$, we see $e^{x+a} = e^a e^x$ for all real x, a.
2. For positive r, s, we write $r = e^x$ and $s = e^a$, to obtain

$$\log(rs) = \log(e^x e^a) = \log(e^{x+a}) = x + a = \log(r) + \log(s).$$

3. Recall that $y = e^x$ satisfies $\frac{dy}{dx} = y$, so that

$$\frac{dx}{dy} = \frac{1}{\frac{dy}{dx}} = \frac{1}{y}.$$

Because $x = \log(y)$, the derivative of $\log(y)$ (with respect to y) is $\frac{1}{y}$.

4. Given $a > 0$ and real x, we define $a^x = e^{x \log(a)}$. By the chain rule, a^x differentiates to

$$\Big(\log(a)\Big)e^{x \log(a)} = \Big(\log(a)\Big)a^x.$$

A trigonometric identity

The derivatives of $\sin(x)$ and $\cos(x)$ are $\cos(x)$ and $-\sin(x)$ respectively, and the derivative of x^2 is $2x$. So, by the chain rule, the derivative of $(\sin(x))^2 + (\cos(x))^2$ is

$$2(\sin(x))(\cos(x)) + 2(\cos(x))(-\sin(x)) = 0.$$

At $x = 0$, $(\sin(x))^2 + (\cos(x))^2 = 0^2 + 1^2 = 1$. Because only constant functions have zero derivative, $(\sin(x))^2 + (\cos(x))^2 = 1$ for all values of x.

Rederiving Madhava's sum

The tangent $\tan(x)$ of an angle x is defined by

$$\tan(x) = \frac{\sin(x)}{\cos(x)},$$

so that $\tan(\pi/4) = 1$ as $\sin(\pi/4) = \cos(\pi/4) = 1/\sqrt{2}$. Now inverse tangent (or arc tangent) $y = \tan^{-1}(x)$ is defined as the value in the range $-\frac{\pi}{2} < y < \frac{\pi}{2}$ such that $\tan(y) = x$. In particular, we have $\tan^{-1}(1) = \pi/4$.

The rules of differentiation applied to the equation $\tan(y) = x$ show

$$\frac{\mathrm{d}y}{\mathrm{d}x} = (\cos(y))^2 = \frac{(\cos(y))^2}{(\cos(y))^2 + (\sin(y))^2} = \frac{1}{1 + (\tan(y))^2} = \frac{1}{1 + x^2}.$$

We found the power series for $1/(1-x)$ in Chapter 3; replacing x with $-x^2$, we find

$$\frac{dy}{dx} = \frac{1}{1+x^2} = 1 - x^2 + x^4 - x^6 + x^8 - x^{10} + \cdots.$$

Noting $y(0) = 0$ and applying term-by-term integration, we finally obtain

$$y = \tan^{-1}(x) = x - \frac{x^3}{3} + \frac{x^5}{5} - \frac{x^7}{7} + \frac{x^9}{9} - \frac{x^{11}}{11} + \cdots.$$

This series is valid for $-1 < x \leqslant 1$ and setting $x = 1$ yields Madhava's sum for π.

e is irrational

We will use Euler's definition for e, namely

$$e = 1 + 1 + \frac{1}{2!} + \frac{1}{3!} + \frac{1}{4!} + \cdots + \frac{1}{n!} + \cdots.$$

So, for any n, we have

$$0 < e - \left(1 + 1 + \frac{1}{2!} + \frac{1}{3!} + \frac{1}{4!} + \cdots + \frac{1}{n!}\right) = \frac{1}{(n+1)!} + \frac{1}{(n+2)!} + \cdots.$$

The right-hand side is less than

$$\frac{1}{(n+1)!}\left(1 + \frac{1}{(n+1)} + \frac{1}{(n+1)^2} + \cdots\right)$$
$$= \frac{1}{(n+1)!}\left(1 - \frac{1}{(n+1)}\right)^{-1} = \frac{1}{n(n!)},$$

using the power series for $1/(1-x)$ found in Chapter 3. Hence

$$0 < e - \left(1 + 1 + \frac{1}{2!} + \frac{1}{3!} + \frac{1}{4!} + \cdots + \frac{1}{n!}\right) < \frac{1}{n(n!)}$$

for any n. If $e = m/n$ were rational, then we could multiply the above by $n!$ to obtain

$$0 < n!\frac{m}{n} - \left(n! + n! + \frac{n!}{2!} + \frac{n!}{3!} + \frac{n!}{4!} + \cdots + 1\right) < \frac{1}{n}.$$

This is a contradiction, as the middle expression is a whole number, but there is no whole number strictly between 0 and $1/n$. The assumption that e is rational leads to a contradiction, and hence e is irrational.

Euler's first solution of the Basel problem

Note firstly that

$$\left(1 - \frac{x}{r_1}\right)\left(1 - \frac{x}{r_2}\right) = 1 - \left(\frac{1}{r_1} + \frac{1}{r_2}\right)x + \left(\frac{1}{r_1 r_2}\right)x^2.$$

The roots of the quadratic on the left are r_1 and r_2, and the sum of their reciprocals equals negative the coefficient of x on the right. And this remains true for finitely many brackets and general polynomials. However, the result is *not* true of power series; for example, the equation $e^x = 0$ has no solutions, yet the coefficient of x in the exponential series is 1.

Nonetheless, Euler noted that

$$\frac{\sin\sqrt{x}}{\sqrt{x}} = 1 - \frac{x}{3!} + \frac{x^2}{5!} - \frac{x^3}{7!} + \cdots,$$

by using the Taylor series for sine. As sine equals zero at $\pi, 2\pi, 3\pi, \ldots$ the above power series equals zero when

$x = \pi^2, (2\pi)^2 = 2^2\pi^2, (3\pi)^2 = 3^2\pi^2, \ldots$ He then 'concluded' that their reciprocals add to negative the coefficient of x yielding

$$\frac{1}{\pi^2} + \frac{1}{2^2\pi^2} + \frac{1}{3^2\pi^2} + \frac{1}{4^2\pi^2} + \frac{1}{5^2\pi^2} + \cdots = -\left(\frac{-1}{3!}\right) = \frac{1}{6}.$$

Rearranging this last equation would give $S = \pi^2/6$. Euler could approximate S by evaluating some partial sums to check the plausibility of his answer, but this provides only supporting evidence and doesn't constitute a rigorous proof.

Chapter 4

Deriving Newton's method

The gradient of the graph $y = f(x)$ at $\left(x_1, f(x_1)\right)$ equals $f'(x_1)$, so the tangent line at this point has equation

$$y - f(x_1) = f'(x_1)(x - x_1).$$

This line crosses the x-axis at $(x_2, 0)$. By substituting $x = x_2$ and $y = 0$ into the equation and solving for x_2, we derive Newton's iteration:

$$x_2 = x_1 - \frac{f(x_1)}{f'(x_1)}.$$

If $f(x)$ is increasing at $x = a$, and the derivative $f'(x)$ is increasing on the interval $a \leqslant x \leqslant x_1$, then the iterations x_n will converge quadratically as a decreasing sequence to the solution a.

The stable Lotka–Volterra equilibrium

At an equilibrium for the Lotka–Volterra equations, we have $F'(t) = 0 = R'(t)$ so that

$$0 = -mF + aFR, \qquad 0 = bR - kFR.$$

There are two solutions to these equations: $(F, R) = (0, 0)$ and $(F, R) = (b/k, m/a)$.

To analyse the second equilibrium, we set

$$F = \frac{b}{k} + \varepsilon_1, \qquad R = \frac{m}{a} + \varepsilon_2,$$

where ε_1 and ε_2 are small enough that we will consider only linear terms involving them. Substituting these expressions back into the Lotka–Volterra equations, we get

$$\varepsilon_1{}' = -m\left(\frac{b}{k} + \varepsilon_1\right) + a\left(\frac{b}{k} + \varepsilon_1\right)\left(\frac{m}{a} + \varepsilon_2\right) = \frac{ab}{k}\varepsilon_2.$$

This last equality is found by expanding the brackets and ignoring the negligible term $a\varepsilon_1\varepsilon_2$. A similar calculation for $\varepsilon_2{}'$ gives $\varepsilon_2' = -(mk/a)\varepsilon_1$. Combining these equations, we have

$$\varepsilon_1'' = (\varepsilon_1')' = \left(\frac{ab}{k}\varepsilon_2\right)' = \frac{ab}{k}\varepsilon_2' = \left(\frac{ab}{k}\right)\left(-\frac{mk}{a}\right)\varepsilon_1 = -mb\varepsilon_1.$$

$\varepsilon_1 = A \sin\sqrt{mb}t + B \cos\sqrt{mb}t$ for constants A and B is the general solution to this differential equation and we can use $\varepsilon_2 = (k/ab)\varepsilon_1{}'$ to find a similar expression for ε_2. These expressions for ε_1 and ε_2 describe small oval (specifically elliptical) orbits around the equilibrium point, showing it to be stable.

Chapter 5

Minimizing the least-squares error

If we expand the brackets in the expression for the total error $E(a, b)$ from Chapter 4, with n data points (rather than 6), we find that

$$E(a, b) = Sa^2 + 2Tab + nb^2 - 2Aa - 2Bb + C,$$

where

$$S = x_1^2 + x_2^2 + \cdots + x_n^2, \quad T = x_1 + x_2 + \cdots + x_n$$

$$A = x_1y_1 + x_2y_2 + \cdots + x_ny_n, \quad B = y_1 + y_2 + \cdots + y_n,$$

$$C = y_1^2 + y_2^2 + \cdots + y_n^2.$$

Now $E(a, b)$ is minimal when $\partial E/\partial a = 0 = \partial E/\partial b$. Differentiating, we find that

$$\frac{\partial E}{\partial a} = 2Sa + 2Tb - 2A = 0, \qquad \frac{\partial E}{\partial b} = 2Ta + 2nb - 2B = 0.$$

These simultaneous equations can then be solved to find the optimal values for a and b.

Lines are the shortest curves

In this example $F(x, f, f') = \sqrt{1 + (f')^2}$. Note F does not depend on f, and so $\partial F/\partial f = 0$. By the chain rule, we have that

$$\frac{\partial F}{\partial f'} = \frac{1}{2}\left(1 + (f')^2\right)^{-1/2} \times (2f') = \frac{f'}{\sqrt{1 + (f')^2}}.$$

So the Euler–Lagrange equation now reads

$$\frac{\mathrm{d}}{\mathrm{d}x}\left(\frac{f'}{\sqrt{1 + (f')^2}}\right) = 0.$$

Finally, $f(x) = x$ solves this as $f'(x) = 1$, so that the expression in the brackets equals $1/\sqrt{2}$; this is a constant, meaning its derivative is zero and so equals the right-hand side.

The curl of a conservative field is zero

Given a vector field $\mathbf{v} = (v_x, v_y, v_z)$, curl($\mathbf{v}$) is the vector field defined by

$$\operatorname{curl}(\mathbf{v}) = \left(\frac{\partial v_z}{\partial y} - \frac{\partial v_y}{\partial z}, \frac{\partial v_x}{\partial z} - \frac{\partial v_z}{\partial x}, \frac{\partial v_y}{\partial x} - \frac{\partial v_x}{\partial y}\right).$$

If $\mathbf{v} = \operatorname{grad}(f) = (\partial f/\partial x, \partial f/\partial y, \partial f/\partial z)$, then

$$\operatorname{curl}(\mathbf{v}) = \left(\frac{\partial^2 f}{\partial y \partial z} - \frac{\partial^2 f}{\partial z \partial y}, \frac{\partial^2 f}{\partial z \partial x} - \frac{\partial^2 f}{\partial x \partial z}, \frac{\partial^2 f}{\partial x \partial y} - \frac{\partial^2 f}{\partial y \partial x}\right) = (0, 0, 0)$$

because, in each case, the order of differentiation does not matter.

Stokes' theorem implies Green's theorem

Green's theorem states that for a region R in the xy-plane, and functions $P(x, y)$ and $Q(x, y)$,

$$\int_{\partial R} P \mathrm{d}x + Q \mathrm{d}y = \iint_R \left(\frac{\partial Q}{\partial x} - \frac{\partial P}{\partial y}\right) \mathrm{d}A.$$

Green's theorem is a special case of Stokes' theorem as follows. Set $\mathbf{v} = \big(P(x, y), Q(x, y), 0\big)$. Because R lies in the plane, the unit normal is $(0, 0, 1)$, so the component of curl($\mathbf{v}$) in this normal direction is just the z-component of curl($\mathbf{v}$). Setting $v_x = P$ and $v_y = Q$ in the above definition for curl, we find

$$\iint_R \operatorname{curl}(\mathbf{v}) \cdot \mathrm{d}\mathbf{S} = \iint_R \left(\frac{\partial Q}{\partial x} - \frac{\partial P}{\partial y}\right) \mathrm{d}A.$$

On the other hand, $\mathrm{d}\mathbf{r} = (\mathrm{d}x, \mathrm{d}y, 0)$ is an infinitesimal tangent vector to the boundary ∂R and

$$\mathbf{v}\cdot\mathrm{d}\mathbf{r} = (P, Q, 0)\cdot(\mathrm{d}x, \mathrm{d}y, 0) = P\mathrm{d}x + Q\mathrm{d}y.$$

Putting these expressions into Stokes' theorem gives Green's theorem.

Chapter 6

Deriving the Fourier coefficients

Setting $t = 0$ in the expression for $y(x, t)$ as an infinite sum of $y_n(x, t)$, we get

$$y(x, 0) = f(x) = A_1 \sin\left(\frac{\pi x}{L}\right) + A_2 \sin\left(\frac{2\pi x}{L}\right) + A_3 \sin\left(\frac{3\pi x}{L}\right) + \cdots.$$

The trigonometric identity

$$2\sin\left(\frac{m\pi x}{L}\right)\sin\left(\frac{n\pi x}{L}\right) = \cos\left(\frac{(m-n)\pi x}{L}\right) - \cos\left(\frac{(m+n)\pi x}{L}\right)$$

can be used to show that

$$\int_0^L \sin\left(\frac{m\pi x}{L}\right)\sin\left(\frac{n\pi x}{L}\right)\mathrm{d}x = 0,$$

when m and n are distinct. And when $m = n$, this integral becomes

$$\frac{1}{2}\int_0^L\left(1 - \cos\left(\frac{2n\pi x}{L}\right)\right)\,\mathrm{d}x = \frac{L}{2}.$$

If we multiply the infinite sum for $y(x, 0)$ by $\sin(n\pi x/L)$ and integrate between $x = 0$ and $x = L$, then all the integrals equal zero, except the nth one. Specifically, we get

$$\int_0^L f(x)\sin\left(\frac{n\pi x}{L}\right)\,\mathrm{d}x = A_n\frac{L}{2},$$

giving Fourier's expression for A_n.

Deriving the Cauchy–Riemann equations

Let $f(z) = u(x, y) + v(x, y)i$ be a complex differentiable function where $z = x + yi$. Then

$$f'(z) = \text{the limit of } \frac{f(z+h) - f(z)}{h} \text{ as } h \text{ becomes small}$$

exists and gives the same value *however* h converges to zero. In the case when h is real, then $z + h = (x + h) + yi$ and so

$$\frac{f(z+h) - f(z)}{h} = \frac{[u(x+h, y) + v(x+h, y)i] - [u(x, y) + v(x, y)i]}{h}$$

$$= \frac{[u(x+h, y) - u(x, y)]}{h} + \frac{[v(x+h, y) - v(x, y)]}{h} i,$$

which has limit

$$\frac{\partial u}{\partial x} + \frac{\partial v}{\partial x} i$$

as h becomes small.

If instead $h = ik$ is purely imaginary, then k becomes small when h becomes small; also note that $z + h = x + i(y + k)$. So

$$\frac{f(z+h) - f(z)}{h} = \frac{[u(x, y+k) + v(x, y+k)i] - [u(x, y) + v(x, y)i]}{ki}$$

$$= \frac{[v(x, y+k) - v(x, y)]}{k} - \frac{[u(x, y+k) - u(x, y)]}{k} i,$$

which has limit

$$\frac{\partial v}{\partial y} - \frac{\partial u}{\partial y} i$$

as h, and so k, becomes small.

As these two limits must be equal, their real parts are equal and separately their imaginary parts are equal, so we've proved the Cauchy–Riemann equations

$$\frac{\partial u}{\partial x} = \frac{\partial v}{\partial y}, \quad \frac{\partial v}{\partial x} = -\frac{\partial u}{\partial y}.$$

Proving Cauchy's theorem from Green's theorem

Let $f(x + yi) = u(x, y) + iv(x, y)$ be a complex differentiable function with real part $u(x, y)$ and imaginary part $v(x, y)$ and let C be an anticlockwise loop in the complex plane. Cauchy's theorem states that

$$\int_C f(z)\ \mathrm{d}z = 0.$$

This can be deduced from Green's theorem. Recall that Green's theorem states that

$$\int_C P\mathrm{d}x + Q\mathrm{d}y = \iint_R \left(\frac{\partial Q}{\partial x} - \frac{\partial P}{\partial y} \right)\ \mathrm{d}x\ \mathrm{d}y,$$

where R is the region bounded by C, and $P(x, y)$ and $Q(x, y)$ are real differentiable functions. Further

$$f(z)\ \mathrm{d}z = (u + iv)(\mathrm{d}x + i\mathrm{d}y) = (u\ \mathrm{d}x - v\ \mathrm{d}y) + i(v\ \mathrm{d}x + u\ \mathrm{d}y).$$

If we apply Green's theorem separately to the real and imaginary parts of $f(z)\mathrm{d}z$, we get

$$\int_C f(z)\,\mathrm{d}z = \iint_R \left(-\frac{\partial v}{\partial x} - \frac{\partial u}{\partial y}\right)\mathrm{d}x\,\mathrm{d}y + i\iint_R \left(\frac{\partial u}{\partial x} - \frac{\partial v}{\partial y}\right)\mathrm{d}x\,\mathrm{d}y$$

and both integrals are zero from the Cauchy–Riemann equations.

Complex logarithm and powers

An important consequence of Euler's identity is that the complex exponential function is periodic. As sine and cosine have period 2π, the exponential function has period $2\pi i$, meaning $e^{z+2\pi i} = e^z$ for any z. This is very different behaviour compared with the real numbers (Chapter 3).

The complex exponential attains all possible outputs except zero; so for any non-zero complex number z, there is a complex number w such that

$$e^w = z.$$

For z, a positive real number, there is a *unique* real number w such that $e^w = z$ and w is called the **logarithm** of z, written $\log(z)$. By contrast, with complex numbers, if w solves $e^w = z$, then so does $w + 2\pi i$, as do $w + 4\pi i$ and $w - 2\pi i$. In fact, there are infinitely many values w that solve the equation. Which of these should we think of as *the* logarithm of z?

Objectively, none of these w is a preferred solution for $e^w = z$, so mathematicians get around this problem by making a choice—a so-called *principal value*—and then working consistently with that choice. In this way it is possible to define a choice of complex logarithm $\log(z)$ which is a complex differentiable function and has derivative $1/z$.

This issue with the complex logarithm also applies to powers of complex numbers. For example, there is no single, preferred value we might assign to i^i. From Euler's identity we know that

$$e^{i\pi/2} = \cos(\pi/2) + i \sin(\pi/2) = 0 + i \times 1 = i,$$

and so we might argue that

$$i^i = (e^{i\pi/2})^i = e^{i^2\pi/2} = e^{-\pi/2},$$

and this is indeed a possible correct answer. But it is also the case that $e^{5i\pi/2} = e^{-3i\pi/2} = i$, due to the periodicity of the exponential function, and these lead to different values for i^i. In fact, there are infinite possible values for i^i; perhaps surprisingly, all of them are real. But again it is possible, for any complex number a, to define a preferred complex differentiable function z^a which has derivative az^{a-1}.

An appreciation of the complex logarithm gives further insight into why $\int_C z^{-1}\,dz$ equals $2\pi i$, where C denotes the anticlockwise unit circle. $n = -1$ is a special case, because in each other case z^n has antiderivative $\frac{z^{n+1}}{n+1}$. This doesn't make sense when $n = -1$, but we know that z^{-1} has antiderivative $\log(z)$.

But we can't smoothly choose principal values for $\log(z)$ on all of C. If instead we think of C as starting at $a = 1$, circling the origin and returning to a *distinct* end point $b = 1$, then we can define $\log(z)$ for each point of C (Figure 67).

Recall that $\log(z)$ is a value w such that $e^w = z$. At $a = 1$ we can set $\log(a) = 0$, as $e^0 = 1$. And at $z = e^{i\theta}$ we set $\log(z) = i\theta$. As we move anticlockwise around C, this choice of $\log(z)$ changes smoothly; when we get to the top of C we have $\log(i) = i\pi/2$, and half-way around C we find $\log(-1) = i\pi$. But as we continue around the lower semi-circle of C, we see that θ is approaching 2π and $\log(z)$ is approaching $2\pi i$. Even though $a = 1 = b$, we assign $\log(b) = 2\pi i$, as b is at the 'end' of C. By the fundamental theorem,

$$\int_C z^{-1}\,dz = \log(b) - \log(a) = 2\pi i - 0 = 2\pi i.$$

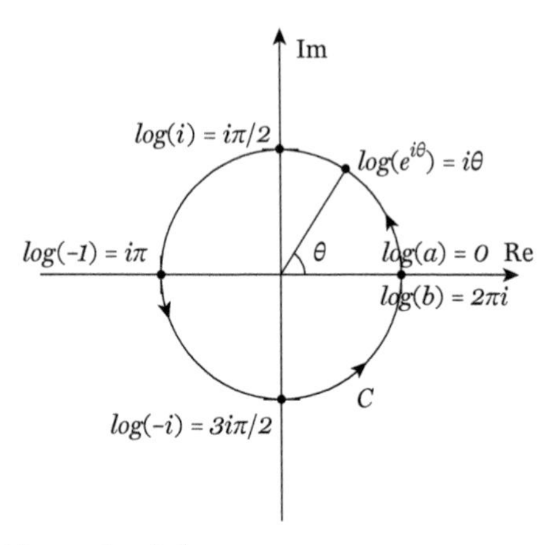

67. Logarithm on the circle.

Put another way, the above integral takes this value precisely because it's the period of the exponential function.

Evaluating an integral with the residue theorem

The first integral listed in Chapter 7 is

$$\int_0^{2\pi} \frac{\mathrm{d}x}{5 - 4\cos x}.$$

To turn this into a complex integral, we set $z = e^{ix}$. As x increases from 0 to 2π, then z moves once anticlockwise around the unit circle C. Now $\mathrm{d}z = ie^{ix}\,\mathrm{d}x = iz\mathrm{d}x$ and, by Euler's identity,

$$\cos x = \frac{e^{ix} + e^{-ix}}{2} = \frac{z + z^{-1}}{2}.$$

Substituting these expressions for $\mathrm{d}x$ and $\cos x$ into the integral, it rearranges to the complex integral

$$\int_C \frac{i\,\mathrm{d}z}{2z^2 - 5z + 2} = \int_C \frac{i\,\mathrm{d}z}{2(z-2)\left(z-\frac{1}{2}\right)}.$$

The integrand here has a singularity $z = \frac{1}{2}$ which, importantly, is inside C and one at $z = 2$ which is outside of C. The residue theorem shows the integral equals $2\pi i \times \left(\text{residue at } \frac{1}{2}\right)$. Finally, the Laurent series centred at $\frac{1}{2}$ begins

$$\frac{i}{2(z-2)\left(z-\frac{1}{2}\right)} = -\frac{i}{3}\left(z - \frac{1}{2}\right)^{-1} - \frac{2i}{9} - \frac{4i}{27}\left(z - \frac{1}{2}\right) - \cdots$$

and so the residue at $\frac{1}{2}$ equals $-\frac{i}{3}$. Hence the integral equals

$$2\pi i \times \left(\text{residue at } \frac{1}{2}\right) = 2\pi i \times \left(\frac{-i}{3}\right) = \frac{2\pi}{3}.$$

Appendix

The real and imaginary parts satisfy Laplace's equation

If $f(x+yi) = u(x,y) + iv(x,y)$ is complex differentiable, the Cauchy–Riemann equations hold—that is,

$$\frac{\partial u}{\partial x} = \frac{\partial v}{\partial y}, \qquad \frac{\partial v}{\partial x} = -\frac{\partial u}{\partial y}.$$

The order of differentiation in mixed derivatives does not matter, so

$$\frac{\partial^2 u}{\partial x^2} = \frac{\partial}{\partial x}\left(\frac{\partial u}{\partial x}\right) = \frac{\partial}{\partial x}\left(\frac{\partial v}{\partial y}\right) = \frac{\partial}{\partial y}\left(\frac{\partial v}{\partial x}\right) = \frac{\partial}{\partial y}\left(-\frac{\partial u}{\partial y}\right) = -\frac{\partial^2 u}{\partial y^2},$$

showing u satisfies Laplace's equation. The same result for v follows similarly.

Chapter 8

A countable set is null

The elements of a countable set can be listed as $r_1, r_2, r_3, \ldots$ Given any positive ε we can cover each r_k with the interval centred on r_k and having length $\varepsilon/2^k$. The total length of these intervals is

$$\frac{\varepsilon}{2}+\frac{\varepsilon}{2^2}+\frac{\varepsilon}{2^3}+\frac{\varepsilon}{2^4}+\frac{\varepsilon}{2^5}+\cdots=\varepsilon,$$

as shown in Chapter 1.

There are infinitely many primes

Euclid proved in his *Elements* that there are infinitely many prime numbers. The proof does not involve analysis but is a classic of mathematics, and so is included here. Let $p_1, p_2, \cdots, p_n$ be finitely many prime numbers. Euclid then introduced the number

$$N = p_1 \times p_2 \times \cdots \times p_n + 1.$$

Note that none of $p_1, p_2, \ldots, p_n$ divides N, because each leaves remainder 1.

There are two alternatives—either N is itself a new prime (as arises from $2 \times 3 \times 5 + 1 = 31$) or the prime factors of N are new prime numbers (as arises with $2 \times 3 \times 5 \times 7 \times 11 \times 13 + 1 = 30031 = 59 \times 509$). In either case, we have found a new prime number not on our list, and this process can be continued indefinitely.

Historical timeline

c.370 BCE Eudoxus develops *Method of Exhaustion* for calculating areas

c.270 BCE Archimedes shows $3\frac{10}{71} < \pi < 3\frac{1}{7}$

c.1400 Madhava and the Kerala school develop power series

1614 Napier introduces logarithms

c.1620 Harriot investigates continuous compounding and the exponential function

1636 Fermat's method of tangents

1637 Descartes introduces Cartesian co-ordinates

1668 James Gregory gives the first proof of a limited version of the Fundamental Theorem of Calculus

1669 Newton shares his *De Analysi* in a limited way

1673 Leibniz coins the term 'function'

1684 Leibniz's first publication on calculus

1687 Newton publishes his *Principia*

1696 Johann Bernoulli poses the brachistochrone problem

1703 Guido Grandi studies the sum $1 - 1 + 1 - 1 + \cdots$

1715 Brook Taylor publishes on Taylor series

1734 Bishop Berkeley's *The Analyst*

1737 Euler shows that e is irrational

1747 D'Alembert derives the wave equation

1748 Euler's Identity in *Introductio in Analysin Infinitorum*

1757	D'Alembert first writes down the Cauchy–Riemann equations. Later developed by Euler (1797), Cauchy (1814), and Riemann (1851)
1760	Lagrange poses Plateau's problem
1761	Johann Lambert proves that π is irrational
1768	Euler's method appears in *Institutionum Calculi Integralis*
1799	Caspar Wessel first represents complex numbers in a plane
1801	Gauss uses least-squares method to relocate the asteroid Ceres
1813	Argand publishes a proof of the *Fundamental Theorem of Algebra* (Gauss is often cited as having proved the theorem earlier, in 1799, but his proof was incomplete, with a significant topological gap)
1817	Bolzano defines convergence and continuity and proves the Intermediate Value Theorem
1821	Cauchy's *Cours d'Analyse* discusses limits and attempts to formalize analysis
1822	Fourier's *Analytical Theory of Heat* introduces Fourier series
1825	Cauchy publishes his integral theorem
1826	Ostrogradsky proves the divergence theorem
1828	Green's theorem appears in *An Essay on the Application of Mathematical Analysis to the Theories of Electricity and Magnetism*
1829	Dirichlet's memoir on the convergence of Fourier series
1829	Dirichlet constructs a function which isn't Riemann integrable
1837	Dirichlet founds analytic number theory
1837	Dirichlet shows an absolutely convergent infinite sum always rearranges to the same limit
1843	Laurent's theorem
1850	William Thomson (Lord Kelvin) writes to Stokes with what is now referred to as Stokes' theorem
1853	Riemann proves his rearrangement theorem
1854	Riemann defines his theory of integration (published 1868)

1861	Weierstrass lectures on $\varepsilon - \delta$ analysis
1870	Weierstrass proves the isoperimetric inequality
1872	Weierstrass describes a continuous function that is differentiable nowhere
1873	Continuous function found with Fourier series that doesn't converge
1873	Hermite shows that e is transcendental
1874	Cantor proves that the real numbers are uncountable
1875	Darboux reformulates Riemann's integral
1882	Ferdinand von Lindemann proves that π is transcendental
1890	Cesàro generalizes the notion of convergence of infinite sums
1896	Prime number theorem proved using analytic methods
1899	J. Willard Gibbs' 'phenomenon' of poor Fourier series convergence near discontinuities
1900	Lyapunov proves a general version of the central limit theorem
1902	Lebesgue defines his theory of integration
1905	Vitali gives an example of a non-measurable set
1910	Ramanujan's approximation to π
1910	Joukowski studies aerofoils using complex analysis
1913	Ramanujan further generalizes the notion of convergence of infinite sums
1918	Hausdorff measure and Hausdorff (fractal) dimension defined
1923	Wiener models continuous-time Brownian motion
1924	Banach and Tarski publish their paradox
1925	Lotka and Volterra (1926) model predator–prey dynamics
1930	Dirac introduces his delta function in *The Principles of Quantum Mechanics*
1932	Von Neumann's *Mathematical Foundations of Quantum Mechanics*
1932	Banach publishes *Theory of Linear Operations*, a seminal work in functional analysis

1936	Sobolev first introduces distributions
1945	Élie Cartan generalizes Stokes' theorem
1950	Schwartz wins a Fields Medal for his work on distributions
1966	Abraham Robinson publishes *Non-standard Analysis*, introducing the hyperreals
1970	Osserman completes proof of Plateau's problem
1992	Different, homophonically indistinguishable drums found
2004	Ben Green and Terence Tao prove their theorem

References and further reading

References

Chapter 2: All change . . . the calculus of Fermat, Newton, and Leibniz

C. H. Edwards Jr, *The Historical Development of the Calculus* (2013) Springer, p. v.

Michael Sean Mahoney, *The Mathematical Career of Pierre de Fermat, 1601–1665* (1994) Princeton, p. 365.

Chapter 3: To the limit: analysis in the 18th and 19th centuries

Richard Earl, *Topology: A Very Short Introduction* (2019) Oxford, Chapters 3 and 4.

Chapter 5: Dimensions aplenty

Richard Earl, *Topology: A Very Short Introduction* (2019) Oxford, Chapter 5.

Chapter 6: I'll name that tune in . . .

Alain Goriely, *Applied Mathematics: A Very Short Introduction* (2018) Oxford.

Chapter 8: But there's more . . .

Morris Kline, *Mathematical Thought from Ancient to Modern Times* (1972) Oxford, p. 655.

Further reading

David Acheson, *From Calculus to Chaos* (1998) Oxford
David Acheson, *The Calculus Story* (2017) Oxford
Lara Alcock, *How to Think about Analysis* (2014) Oxford
Carl B. Boyer, *The History of the Calculus and Its Conceptual Development* (1959) Dover
Ellen Flower, 'The "Analysis" of a Century' (2021), winner of the British Society for the History of Mathematics undergraduate essay prize: https://people.maths.ox.ac.uk/earl/
Ivor Grattan-Guinness (ed.), *From the Calculus to Set Theory 1630–1910* (1980) Princeton
Eli Maor, *e: The Story of a Number* (1998) Princeton
Leonard Smith, *Chaos: A Very Short Introduction* (2007) Oxford
John Stillwell, *The Real Numbers: An Introduction to Set Theory and Analysis* (2013) Springer
Glenn Van Brummelen, *Trigonometry: A Very Short Introduction* (2020) Oxford

Index

E

F

G

H

I

J

K

L

M

N

P

Q

R

S

T

U

V

W

NUMBERS

A Very Short Introduction

Peter M. Higgins

Numbers are integral to our everyday lives and feature in everything we do. In this *Very Short Introduction* Peter M. Higgins, the renowned mathematics writer unravels the world of numbers; demonstrating its richness, and providing a comprehensive view of the idea of the number. Higgins paints a picture of the number world, considering how the modern number system matured over centuries. Explaining the various number types and showing how they behave, he introduces key concepts such as integers, fractions, real numbers, and imaginary numbers. By approaching the topic in a non-technical way and emphasising the basic principles and interactions of numbers with mathematics and science, Higgins also demonstrates the practical interactions and modern applications, such as encryption of confidential data on the internet.

www.oup.com/vsi

STATISTICS

A Very Short Introduction

David J. Hand

Modern statistics is very different from the dry and dusty discipline of the popular imagination. In its place is an exciting subject which uses deep theory and powerful software tools to shed light and enable understanding. And it sheds this light on all aspects of our lives, enabling astronomers to explore the origins of the universe, archaeologists to investigate ancient civilisations, governments to understand how to benefit and improve society, and businesses to learn how best to provide goods and services. Aimed at readers with no prior mathematical knowledge, this *Very Short Introduction* explores and explains how statistics work, and how we can decipher them.

www.oup.com/vsi

INFORMATION

A Very Short Introduction

Luciano Floridi

Luciano Floridi, a philosopher of information, cuts across many subjects, from a brief look at the mathematical roots of information - its definition and measurement in 'bits'- to its role in genetics (we are information), and its social meaning and value. He ends by considering the ethics of information, including issues of ownership, privacy, and accessibility; copyright and open source. For those unfamiliar with its precise meaning and wide applicability as a philosophical concept, 'information' may seem a bland or mundane topic. Those who have studied some science or philosophy or sociology will already be aware of its centrality and richness. But for all readers, whether from the humanities or sciences, Floridi gives a fascinating and inspirational introduction to this most fundamental of ideas.

'Splendidly pellucid.'

Steven Poole, The Guardian

www.oup.com/vsi

CHAOS

A Very Short Introduction

Leonard Smith

Our growing understanding of Chaos Theory is having fascinating applications in the real world - from technology to global warming, politics, human behaviour, and even gambling on the stock market. Leonard Smith shows that we all have an intuitive understanding of chaotic systems. He uses accessible maths and physics (replacing complex equations with simple examples like pendulums, railway lines, and tossing coins) to explain the theory, and points to numerous examples in philosophy and literature (Edgar Allen Poe, Chang-Tzu, Arthur Conan Doyle) that illuminate the problems. The beauty of fractal patterns and their relation to chaos, as well as the history of chaos, and its uses in the real world and implications for the philosophy of science are all discussed in this *Very Short Introduction.*

'. . . Chaos . . . will give you the clearest (but not too painful idea) of the maths involved . . . There's a lot packed into this little book, and for such a technical exploration it's surprisingly readable and enjoyable - I really wanted to keep turning the pages. Smith also has some excellent words of wisdom about common misunderstandings of chaos theory . . .'

popularscience.co.uk

www.oup.com/vsi